The Unofficial Guide to Japanese and International Transformers™

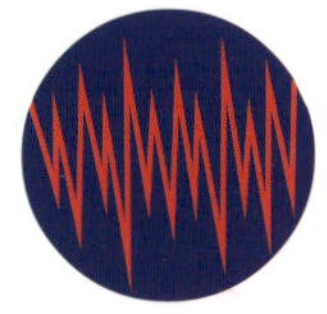
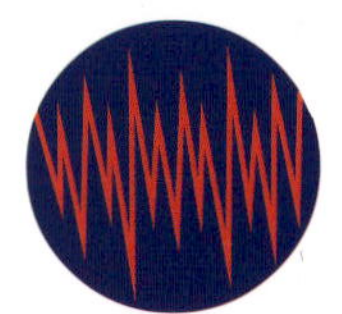

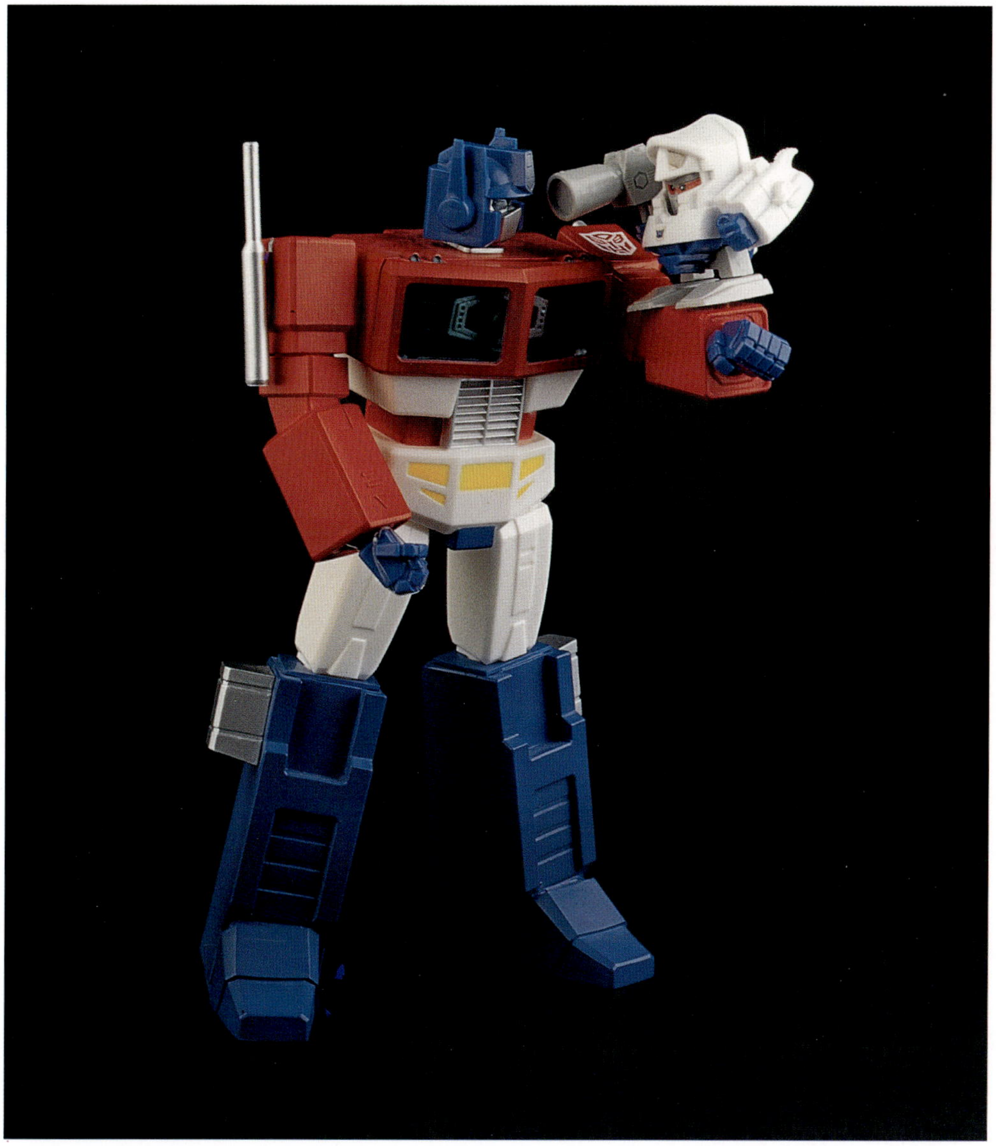

J.E. Alvarez

880 Lower Valley Road, Atglen, PA 19310 USA

Dedication

This one goes out to all my boys:

Fran (Rumble250) O'Boyle, Ron (Sonic Bomber) Bahr, Scott (The Non-believer) Talis, Rob T. Yee, and to the ducks on Pleasant Valley Road.

Disclaimer:

The Transformers, and related characters and elements, are registered trademarks of Hasbro Inc. and Takara Co. Ltd.

The characters and other graphics depicted in this book are copyrighted. None of these copyright holders authorized this book, furnished, nor approved any of the information contained therein. This book in no way attempts to infringe on any copyright, and is based on the author's independent research.

Library of Congress Card Number: 00-109376

Designed by John P. Cheek
Type set in Bedrock/Humanist 521 BT

ISBN: 0-7643-1282-0
Printed in China
1 2 3 4

Published by Schiffer Publishing Ltd.
4880 Lower Valley Road
Atglen, PA 19310
Phone: (610) 593-1777; Fax: (610) 593-2002
E-mail: Schifferbk@aol.com
Please visit our web site catalog at
www.schifferbooks.com or write for a free catalog.

We are always looking for authors to write books on new and related subjects. If you have an idea for a book, please contact us at the above address.

This book may be purchased from the publisher.
Please include $3.95 for shipping.

In Europe, Schiffer books are distributed by
Bushwood Books
6 Marksbury Ave.
Kew Gardens
Surrey TW9 4JF England
Phone: 44 (0)20-8392-8585; Fax: 44 (0)20-8392-9876
E-mail: Bushwd@aol.com
Free postage in the UK. Europe: Air mail at cost.
Please try your bookstore first.

Contents

Acknowledgments

Well, Transfans, welcome back! After the success of *The Unofficial Guide To Transformers: 1980s through 1990s* (Schiffer, 1999) the gang at Schiffer thought that I should make history one more time by writing a sequel. Here we are, so let's start off by acknowledging the misguided souls who made this book what it is.

First I need to start off by thanking the other two founding members of Team-Earth-Matrix. The first member, making his grand debut in a J. E. Alvarez book, is my good friend Ronald Bahr. He is the only other person who spends as much time visiting Scott Talis at the Pennsauken Mart as me! He's like the store mascot or something. I blame myself for Ron's addiction to Japanese Transformers. If I hadn't spent so much time talking about them and showing them off then maybe his tragic addiction to them could have been avoided. In any case, I'm glad that he does collect them because I get to put his great collection (or what's left after I bought it) in this book. If you'd like to get in touch with Ron (for what ever reason) e-mail him at rudyb@nothinbut.net.

The other member of Team-Earth-Matrix is Fran O'Boyle. You might have heard of his buddies list at Transfan@onelist.com. Fran is a great friend of mine with a nice Transformers collection. He's what I call a high roller because he spends just as much money on Transformers as I do. Fran was a great help in obtaining many of the rare toys that you'll see in this book. He has gone above and beyond what I've asked of him so that this book could be made. I also feel compelled to mention that his birthday is the day before mine, on April 13th. If you would like to reach him, e-mail him at rumble250@aol.com.

Like every good sequel, there are always returning characters that survive the first story and get to make a second appearance. I need to thank my employer and mentor Scott Talis, the owner of Play With This Collectible Toys and Videos. Not only does he get the honor of having this book dedicated to him, he also gets to see me every Sunday at work. About 65% of my collection came from his store's stock, which includes all types of toys from the '60s, '70s, '80s, '90s, and beyond. Even though he may look scary, he is still a nice guy—when he is not beating me up. So if you're in South Jersey, stop by his store or write to him at Play With This Collectible Toys and Videos, 339 Pennsauken Mart, Pennsauken New Jersey 08110. The number there is 856-486-4556, or e-mail Scott at toys339@aol.com. If you're on eBay™ then check out his auctions under his seller's name, "ibytoys".

Also returning is George Hubert, Jr. George is a friend of mine and was a great help in the making of this book. Just like before, George lent out his collection for me to photograph and destroy. He also helped me with a lot of the pre-Transformers stuff. If you're ever looking for someone to make you a custom Action Master, e-mail him at theweeter@aol.com

Another big influence on how the way this book turned out is my good friend Robert T. Yee. Rob was the nicest guy right from the start. He gave me a lot of good ideas of what this book should be like, as well as giving me a lot of Transformers to photograph. He showed up with box, after box, after box of stuff (and most of it left with other people)! He also hooked me up with a few other contacts, and dedicating part of this book to him is the least I can do to repay him for all his kindness. A lot of the choice stuff in this book belongs to Rob so if you would like to e-mail him regarding anything you see here, do so at Asylum13x@aol.com.

One of the people that Rob hooked me up with was Benson Yee. You've probably heard of his awesome website, BWTF.com. Ben was an advisor for the Beast Wars television series and another one of those people who are just filled with too much knowledge about the Transformers universe. He is another nice guy who let me borrow toys for this book. Make it a point to check out BWTF.com for all the latest news and info.

I need to thank another person who has become a good friend of mine, Jim Walters, who also brought over a ton a stuff to be photographed. He left with more stuff than he walked in with! In fact, Jim taught me a thing or two about spending money on Transformers...I'm really not in his league. If you would like to e-mail Jim you can at Groo215@aol.com or check out his website, the Transformers Highland, at http://transformers.winterchill.com.

Two other people who were a great help creating this book were Peter and Charles Liu. Charles introduced me to a few other collectors who have become good friends of mine. He also helped me out by transforming all his toys for me and keeping his cool during the shoot. Charles is a big Japanese toy collector so e-mail him at aim_less@iname.com.

There are a few other people who helped me in the making of this book. First up I need to thank George Patouhas and Brad Bricker for bringing their stuff over and for helping out with the beer. They are pretty cool guys to hang around with and they love talking Transformers. If you'd like to e-mail George he can be reached at GreeksInSpace@aol.com and Brad at Brickdogg1@aol.com. Next I need to thank Will Cintrón. Mi hermano! This is another guy I have the pleasure of knowing. Even though Will only lent me one thing to photograph for this book you still gotta love him because he is just oozing machismo. Will loves talking about Beast Wars with anybody so e-mail him at maniacal1@netzero.net. One more person I need to thank is Zachary Clark. He was really nice and brought over some cool stuff so e-mail him at RedeemerV@aol.com. I also need to thank Matthew

Camaratta for sending up those model kits from Florida, having never even spoken to me before. Matt has a webpage where he sells Transformers at transformers.homepage.com, or e-mail him at sykopunk@netzero.net. This is the point were I thank the Co-Presidents of my unofficial fan club, Daniel J. Paglione and Julian J. Paglione. Two really nice kids who lent me their stuff to be immortalized in this book. I hope that one day they can grow up to start the *Official* J. E. Alvarez Fan Club! Maybe next time they can let me borrow some more stuff.

I also need to thank my awesome website designer Jeff Cohn for taking the time to put together a very cool website for me. If you want to get in touch with him, check out his website at http//www.geocities.com/iconmakermktg.

As always I gotta send mad props out to my girl Molly Higgins for the countless hours she spent on the phone with me trying to explain things so even I could understand them. She also did a great job of speel chaecking this book and fixing all of me grammaatical erorrs. She also took most of the pictures for this book after driving an hour and worrying about her sick iguana, while at the same time trying to deal with like fifteen guys jumping all over the place going nuts over toys. If there is anything I could say to Molly it would be next time bring some more film... I guess it wouldn't hurt to thank Mr. Peter Schiffer, and all the beautiful people at Schiffer Publishing who worked on this book.

One last person I need to thank is my mom Gioconda Maglione for buying me my very first Transformer at K-Mart in 1984, as well as for putting up with me for twenty one-years. Thanks to her and everyone else who helped me make this book possible including all the people who went out and bought a copy of the first book.

Introduction

In 1984 a company named Hasbro brought a line of toys known as Transformers to the United States from Japan, thus changing the fate of the thousands of people who live in this world.

When Transformers first hit the streets they were a big hit with kids, collectors, and anyone who loved giant robots from outer space. For anyone who grew up in the 1980s, the names Optimus Prime, Megatron, Starscream, and Bumblebee will always bring back fun memories. In the 1990s Transformers continued to make their mark with the popular line called Beast Wars. Spanning three decades, over seven TV shows, and hundreds upon hundreds of figures, their legend cannot be told in just one book. After the release of my first book, *The Unofficial Guide To Transformers: 1980s through 1990s*, I knew that I wanted to put together the story of some of the lesser-known characters. While Transformers flourished in America, Takara was distributing their own figures in Japan. While most of the toys released in Europe and America also made it onto Japanese markets, many figures remained Japanese exclusives. This guide deals mostly with the figures which were never released outside Japan. For the most part, these figures are rare to come by and very expensive to obtain. Some of the characters are not well known outside Japan and for some of the ones that are might have a different name, color scheme, or allegiance. Even their packaging is different as well as the story lines and continuities that the TV shows were based on.

In this book you will be introduced to the Japanese company known as Takara. During the late '70s and early '80s Takara manufactured a series of toy lines which would one day be the inspiration for the Transformers. These were the Diaclone, Diakron, and Microman lines. In fact, not only were they the inspiration for Transformers, but many of the original 1980s Transformers used the same molds as these figures. In some instances the original colors of the toy were kept the same, which lead to several variations on the Transformers side of things—like the Optimus Prime with gray Roller, the blue Bluestreak, and a figure known as Bumblejumper. Included in Chapter One is a small look at several of the different pre-Transformers era toys. In the next two parts of the book we'll explore the world of Japanese and other foreign Transformers. In the final section of this book you will get to see many of the so called "odd ball" stuff. This includes items from all over the Transformers universe such as lunch boxes, comic books, radios, and pencil sets with the Transformers name or logo attached to them. These are items which not every collector looks for, but for the ones who do some of these pieces are just as valuable as the toys themselves.

Also accompanying the pictures in this book are prices for all the toys shown. These prices are an estimate of the value, based on the condition of the items shown and how difficult it is to obtain them. The prices are just to be used as a suggestion and are based on my experiences as a toy dealer and a professional toy collector.

In the text, if a figure's name is different in Japan from that in other parts of the world, the English version of the name will appear next to it in (parentheses).

From America, to Europe, to the far east, these "robots in disguise" have been part of an incredible phenomenon which has no end in sight. Thanks to the Internet, VCRs, comic books, and the love of these toys and characters in the hearts of fans everywhere, Transformers will forever live on as part of our culture.

Transform and roll out!

The Headmasters! I always think that there was a hidden joke in that name. That and Breastforce.

CHAPTER ONE
The Pre-Transformers Era

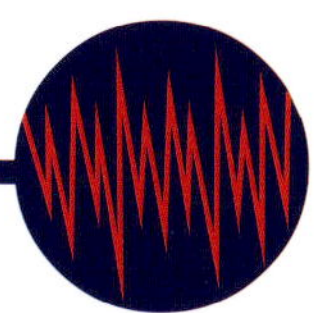

A long, long time ago in a country far away, three lines of transforming robots were brought forth. These were the Diaclone, Diakron, and Microman lines, released by a Japanese toy manufacturer named Takara. These toys incorporated many moving parts, eye-poking missile-shooting weapons, and die-cast metal components. Later on they would become the inspiration for a new line of transforming robots called Transformers.

The most popular of these lines were Takara's Diaclone figures, which made their debut in 1980. It was from this group that Hasbro and Takara came up with most of the early Transformers. The story behind these figures was that the Diaclones were the good guys who would do battle against the evil Warader Empire who came to Earth wanting to take over. Controlling the mighty robot vehicles were the Diaclone Drivers, small human-like figures which could fit inside the robots and pilot them.

Some of the original Transformers which came from this line were the Autobot cars including Optimus Prime, Ultra Magnus, the Omnibots, some of the Triple Changers (like Blitzwing and Astrotrain), and everyone's favorites, the Jumpstarters Twin Twist and Topspin. Other figures that came from this line were the Constructicons, the Trainbots, and the Dinobots. Figures that came from the Warader Empire included the Insecticons Bombshell, Kickback, and Shrapnel. Takara did release some of these figures in America under the names Diakron and Kronoforms.

Also used for the original Transformers concept was another popular toy line from Takara called Microman. These figures first showed up in 1974, and were later renamed New Microman, and then another toy series was added named Microchange. Many Transformers came from this final line including Micro Cassettes Condor (Laserbeak), the Rumble and Frenzy figures, and Jaguar (Ravage). Other characters were the Microbot cars Bumblebee, Cliffjumper, and Bumblejumper. All three of these figures were available in blue, yellow, and red. This is where all the first-assortment Transformers Minicar variations came from. Bumblebee and Cliffjumper were each released as Transformers in both yellow and red. Bumblejumper is the name given to the third figure by Transfans. It was never meant to be released as a Transformer, but some of these figures did make it onto some of the Bumblebee and Cliffjumper backer cards. Other Minicars originally produced by Takara were Brawn, Gears, Huffer, and Windcharger. Some other toys which you might recognize are Camera Robo (Reflector), Cassetteman (Soundwave, who also came with a red version of Rumble), and the predecessors to Blaster, Perceptor, and Megatron. Molds were also taken from two smaller lines of toys that came from the Takatoku company. The first line was Dorvack, where Roadbuster and Whirl came from. The second group was Beetras, which is where the Deluxe Insecticon Transformers had their origins.

The pictures shown in this chapter are just a small glimpse of all the different pre-Transformers items that exist. Many of these figures were issued with different colors, heads, or weapons from that of their Autobot and Decepticon counterparts. Even though many original Transformers were taken from these earlier toy lines many Transformers collectors do not consider them part of the Transformers universe.

This is the Diaclone version of Convoy, named Convoy, which also has the name Diaclone written on the sides of the trailer. *Courtesy of Benson Yee.*

The Walther P-38 U.N.C.L.E. version of Megatron. *Courtesy of Benson Yee.*

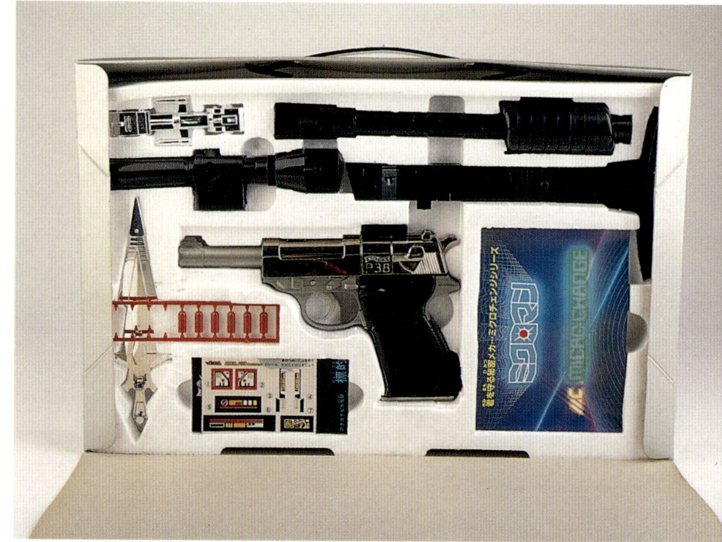

Here we see what the inside of the box looks like.

A similar version of Megatron is the .44 Magnum. *Courtesy of George Hubert.*

The Microman version of Blaster, which is blue like Twincast. *Courtesy of George Hubert.*

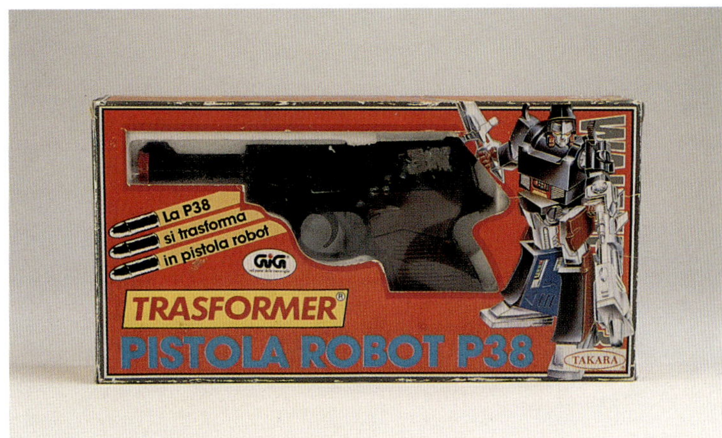

This later version is based on the .44 Magnum design. *Courtesy of George Hubert.*

Recorder Robot later became Soundwave and Soundblaster. Notice the Frenzy we had to slip into the package, the tape that came with this toy was also red. *Courtesy of George Hubert.*

This early Microman version of Frenzy has Takara written across the front of the tape. *Courtesy of George Hubert.*

Here we see red and blue versions of Laserbeak and blue and brown version of Ravage. *Courtesy of George Hubert and George Patouhas.*

Dinosaur Robo is an earlier form of Grimlock that comes with a pilot you can fit into his back while in dino-mode. *Courtesy of George Hubert.*

And here is what the back of the box looks like.

...and in window boxes were you could see the drivers next to the figures. *Courtesy of George Hubert.*

The Powerdashers came packaged both in regular boxes...

The Jumpstarters originated as Diaclone Attackrobos. *Courtesy of George Hubert.*

Here are some of the early Diaclone versions of Bluestreak, which has black going down the center of him in car mode, a yellow Sideswipe, and a black Ratchet. *Courtesy of George Hubert.*

Sometimes the Diaclone versions of Transformers were a different color. Seen here with this blue Convoy are Ultra Magnus, who came with a smaller transforming robot; and Skids, who also came with a small scooter for his Diaclone Driver to ride around in. The scooter can also fit into the back of Skids while in machine mode. This version of Skids also has a different head from the Transformers version. *Courtesy of George Hubert.*

Other color variations included a yellow and a blue Trailbreaker, as well as a red colored Sunstreaker. *Courtesy of George Hubert.*

Sometimes the toy in the box wouldn't even match the color of the toy on the outside, as seen with these versions of Sunstreaker and Trailbreaker, accompanied by Hound and Tracks. *Courtesy of George Hubert and Jim Walters.*

Takara's first version of Inferno. *Courtesy of George Hubert.*

Smokescreen, with Italian and Japanese versions of Bluestreak. *Courtesy of George Patouhas and George Hubert.*

On top is the Japanese version of Mirage, next to the Italian version of Ratchet. On the bottom are the Italian and Japanese versions of Wheeljack. *Courtesy of George Hubert and Jim Walters.*

The European and Japanese versions of Jazz. *Courtesy of Jim Walters and George Hubert.*

The yellow Japanese version of Sideswipe and the red Italian version. *Courtesy of George Hubert and Jim Walters.*

Blue and silver versions of Skids. Notice the change in artwork from the Italian to the Japanese version. *Courtesy of Jim Walters and George Hubert.*

The Diaclone Devastator Gift Set. *Courtesy of George Hubert.*

Two of the original Trainbots Suiken and Kaen. *Courtesy of Fran O'Boyle.*

CHAPTER TWO

Japanese Transformers

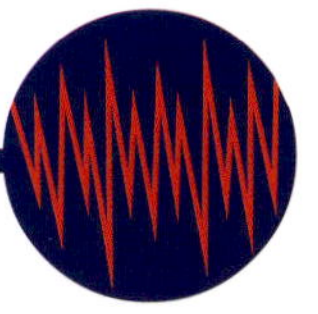

Part 1: Fight! Super Robot Lifeform Transformers!

Soon after Transformers made their 1984 debut in America, Takara was quick to re-introduce the idea of transformable cars and planes to the Japanese market. Dropping the Diaclone and Diakron names, both Takara and Hasbro saw much profit in this new incarnation of these already classic figures. In Japan the Transformers TV show was called *Fight! Super Robot Lifeform Transformers!* These mighty robots came from the planet Seibertron (Cybertron). Instead of Autobots and Decepticons, the two warring factions were the Cybertrons (Autobots) and the Destrons (Decepticons). The Cybertrons were lead by Convoy, who is more commonly referred to as Optimus Prime. Soon after, Transformers became a hit in many parts of the world.

Through 1984 to the end of 1985 the story lines in Japan, Europe, and the United States remained the same. The first major splits in the continuities was the one short episode released exclusively to Japanese video called *Scramble City*. This was intended to be the last episode ever to air before the release of the Transformers movie in 1986. Here the characters of Ultra Magnus, Metroplex, Trypticon, Ratbat, and the Autobot cassettes are introduced. The basic plot for this animated short story encompassed the construction of a new Transformer named Metroplex. Once the Destrons learned of this new weapon the battle breaks out between the Cybertrons and Destrons and doesn't end until Metroplex makes his appearance and saves the day. The show ends with Trypticon rising out of the sea with a mighty roar in the style of Gogira (Godzilla). Unfortunately the second episode to *Scramble City* was never animated, thus letting the story continue in *Transformers: The Movie*.

The original Convoy was the first Transformer released. That's why he has 01 as his designation on the box, $700-1000+. *Courtesy of Robert T. Yee.*

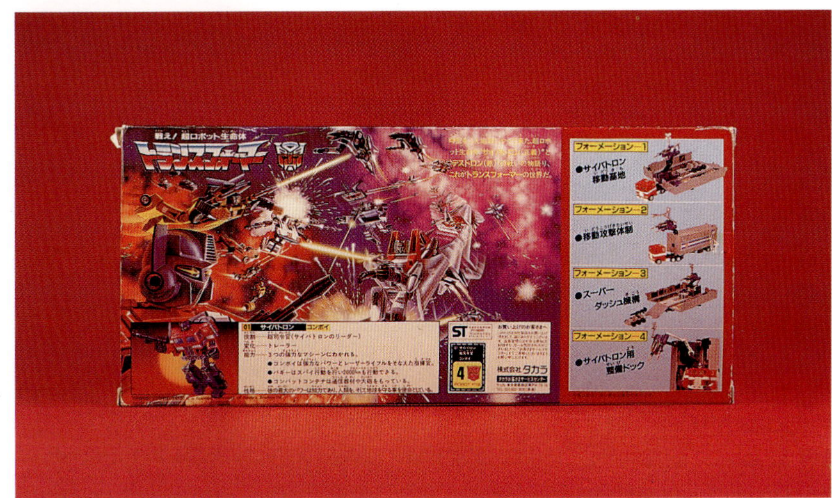

Here is the back of the box, where it's shown that the trailer can stand on its end and work as a repair station.

Jazz was also included in the first assortment of Transformers in Japan, $145-155. *Courtesy of Ron Bahr.*

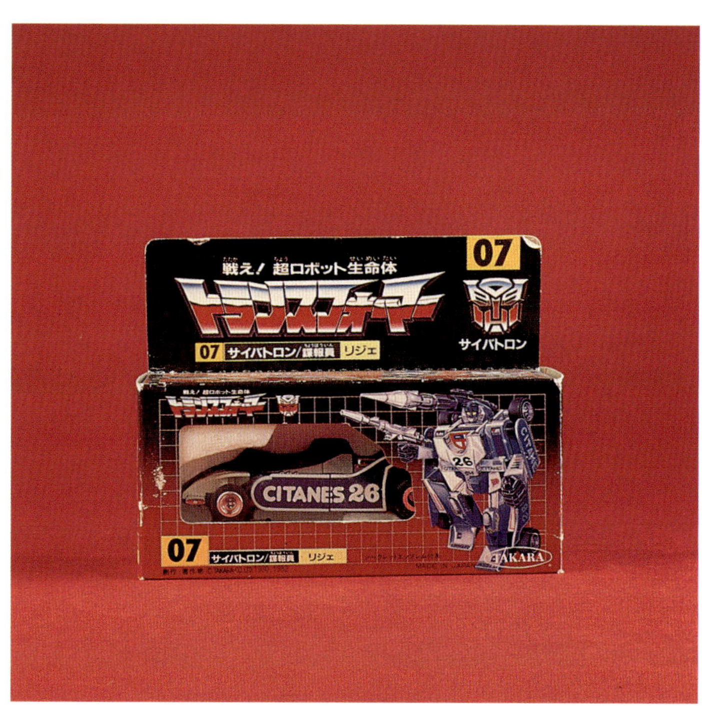

Mirage, $135-150. *Courtesy of Robert T. Yee.*

The initial release of Megatron did not come with the extra gun attachments. It did come with a chrome sword and had blue on the inside of the arms and legs instead of the more common red. The figure can also shoot out pellets. Notice that there is no chrome on the figure itself. $455-500. *Courtesy of Robert T. Yee.*

Rumble $45-50, Soundwave $300-325, and Ravage $60-65. *Courtesy of George Patouhas, Fran O'Boyle, and Ron Bahr.*

Thundercracker, $245-255. *Courtesy of Robert T. Yee.*

Laserbeak was released several times with a different number for each release. Boxed version, which is the first edition, $45-50. Carded second edition, $55-60. *Courtesy of Ron Bahr and Fran O'Boyle.*

This version of Snarl came with black missiles instead of the more common red ones, $200-230. *Courtesy of Ron Bahr.*

Skywarp, $235-250.
Courtesy of Robert T. Yee.

Blaster, $75-95. *Courtesy of Robert T. Yee.*

When Charles mailed this to me to be photographed he specifically told me not to transform it...so I had Rob and Jim do it. I won't tell you what else we did with it!

This is the extremely rare white version of Astrotrain, $450-500+. *Courtesy of Charles Liu.*

As the Military Operations Commander, Shockwave commands $400-450 for you to own him. *Courtesy of Robert T. Yee.*

Here is the white version next to the common gray and purple version of Astrotrain.

The back of Shockwave's box shows an alternate battle scene with Devastator and Bliztwing, instead of the more common one featuring Grimlock and Jetfire.

The white Astrotrain also transforms into a black train instead of a purple one. Loose, this figure can sell for $200-230. *Courtesy of Charles Liu.*

Here is an American Slingshot, carded, with several stickers on it of Chinese writing. *Courtesy of J. E. Alvarez.*

Detail of the front of the card.

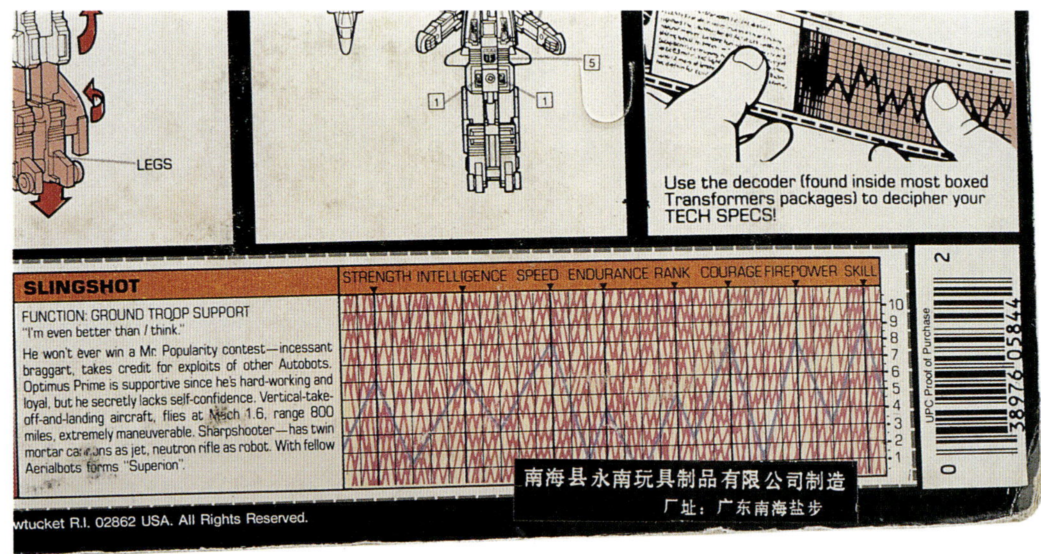

The back of the card also has a sticker which might state the name of a distributor of Transformers in China. Several other carded figures have also been seen with these types of stickers.

After the movie came the next major change. In America and the rest of the world the TV show continued under the name *Transformers*, but in Japan the show was renamed *Transformers 2010*. The show followed the adventures of the new leader of the Cybertrons, Rodimus Convoy (Rodimus Prime). This series was a bit darker than the original show and focused more on the battle in space and on Seibertron, rather than on Earth. The show expanded the roles of the main characters introduced in the movie as well as introduce new characters like Sky Lynx, the Predicons, and the Technobots.

Since Trypticon was introduced in the episode *Scramble City*, his box features the logo for that particular episode, $200-235. *Courtesy of Ron Bahr.*

The back of Trypticon's box shows another common battle sequence.

The same episodes that aired in Japan were shown around the world; however, the story lines split apart after the two-part episode *Double Convoy!* (The Return Of Optimus Prime). As the title suggests, Convoy is resurrected in order to save the galaxy from the horrible hate plague which threatened to destroy all life—Human, Seibertronian, and otherwise. In order to find a cure for the virus, Convoy must take back the *Matrix of Leadership* from Rodimus Convoy, who has been infected with the plague. By the end of 1986, the TV show was losing its popularity throughout the rest of the world. Only three more episodes would air outside of Japan before the show was finally canceled. This was the three-part episode *The Rebirth*, in which the Headmasters and Targetmasters are introduced. These three episodes never aired in Japan, where the show was still bringing in excellent ratings. Instead, as 1987 rolled around, so did a new Transformers TV series entitled *Transformers Headmasters!*

It also shows how both Onslaught and Motormaster can transform into base modes and hook up to Trypticon.

By the time *Transformers 2010* began, the Destrons had a D and their number designation following it, at the same time the Cybertrons had a C and a number following that. From then on for every Destron there was a Cybertron with the same number. Onslaught D-64, $60-65. *Courtesy of Robert T. Yee.*

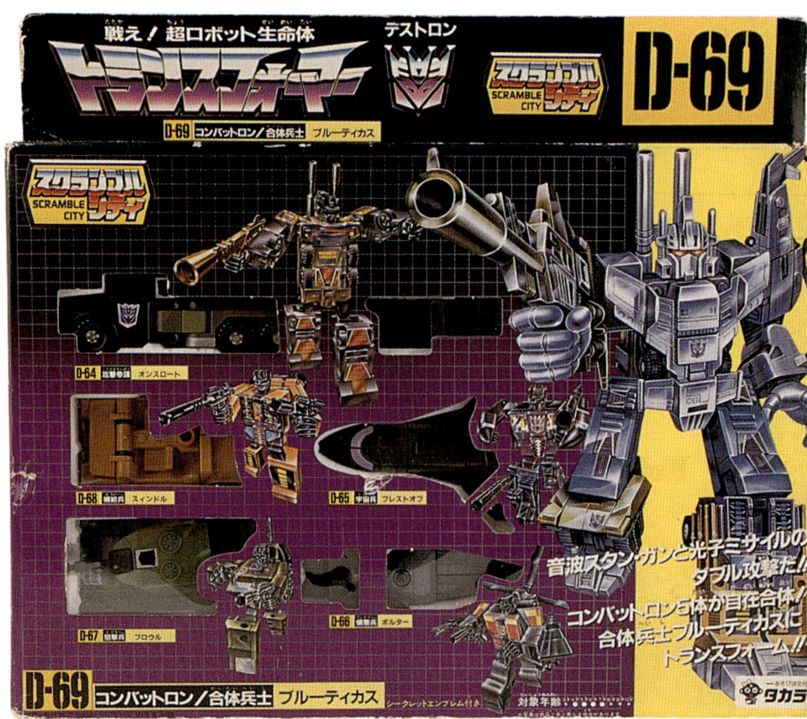

The Bruticus Gift Set was an exclusive to the Japanese market, $400-450. *Courtesy of Robert T. Yee.*

Swindle D-68, $30-35. *Courtesy of Jim Walters.*

Ramhorn C-66, $30-35. *Courtesy of Fran O'Boyle.*

The Predicon Gift Set was also limited to Japanese collectors and is rare, $500-550+. *Courtesy of Charles Liu.*

Scourge, leader of the Sweeps D-71, $200-235. *Courtesy of Ron Bahr.*

The box for the Predicon Gift Set came with a handle so it could be carried around like a suitcase. *Courtesy of Jim Walters.*

The Protectobots Gift Set C-76, $300-330. *Courtesy of Robert T. Yee.*

Blurr $175-200, Kup $185-215+, and Springer $200-225+. *Courtesy of Robert T. Yee.*

Rippersnapper D-80, $25-30. *Courtesy of J. E. Alvarez.*

Wheelie C-80, $25-30. *Courtesy of J. E. Alvarez.*

The Abominus Gift Set is another Japanese exclusive which is very rare and desired by collectors, $440-465+. *Courtesy of Robert T. Yee.*

Transformers Headmasters picked up were *Double Convoy!* left off. In the continuity outside Japan, the Headmasters and Targetmasters had formed an alliance with Humans and Nebulans (a human-like species) to operate inside the Transformers' heads and weapons. However, in Japan the heads were just smaller robots which built the larger robot bodies for transport as well as to transform into vehicles, animals, or robots. The Targetmasters were ordinary Transformers who had small companion robots that could turn into their guns. On the show as well as among the toys, the heads and guns could be interchangeable with any of the bodies in order to gain a special power. Each Headmaster had a special power rating so that when the heads are switched around with another body, that body gained that particular Headmaster's power levels. In the beginning episodes of the show, Convoy dies once again, leaving Hot Rodimus (Hot Rod) to take over and become Rodimus Convoy yet again. Later on in the show Rodimus Convoy, Kup, and Blurr take off in search for a new planet to colonize after the destruction of Seibertron. This leaves Fortress (Cerebros, who could merge with his ship Maximus in order to become Fortress Maximus) in command with Ultra Magnus and Chromedome as his right hand robots.

The Destrons were still being led by Galvatron (who is really Megatron, the original Destron leader in a new body) along with Mega Zarak (Scorponok). Sixshot was, in a sense, second-in-command, with the Destron Headmasters backing up Mega Zarak. Scourge and Cyclonus continue to make appearances throughout the series; however, these characters took on more comic relief or Waspinator-like roles rather than being the vile and evil menaces portrayed in the movie and in *Transformers 2010*. Other characters that showed up during the course of the show were the Twinbots, the Duocons, the Monsterbots, Spike, his wife, and their annoying son Daniel.

Other Transformers re-introduced in this series after their demise were a black version of Soundwave (renamed Soundblaster) and a blue version of Blaster (renamed Twincast). These figures were indentical to their first incarnations except they could now hold two tapes in their chest instead of just one. The Soundblaster and Twincast toys were never released outside of Japan. New tapes were also introduced, but these figures never made it onto any of the television shows. Both sets of tapes were Cybertrons and could transform from a tape to a dinosaur, and could then merge into a larger robot. The first set of these were Gurafi (a pterosaur) and Noizu (a tyrannosaurus) who could join together to form Decibel. With this first set each figure can sell for $250 to $300+ apiece. The second harder-to-find set of tapes included Dairu (a brontosaurus) and Zauru (a lizard). These figures combined into Legout. These two can sell for $450 to $500+ apiece loose and close to $1000 if in the box. These last two Transformers are some of the hardest-to-get figures out of the entire line.

There were also two figures which rank with Legout in that hardest-to-find category. These are the two Targetmasters Artfire (also reffered to as Autofire) and Stepper. Both of these figures were classic Cybertronian cars repainted with some retooling in order to hold a Targetmaster robot. Artfire was a red and white version of Inferno who came with Nightstick; outside Japan the figure was known as Fracus and came originally with Targetmaster Scourge. Stepper was a black version of Jazz who came with Nebulon; the figure was also known as Nightstick who originally came with Cyclonus.

During this series Takara also released individual Headmaster heads. These figures were known as the Headmaster Warriors. This was a great idea on Takara's part, because if you lost the Headmaster to your robot, it wouldn't have a head. Unfortunately these six figures are hard to come by and are very, very expensive. The three Headmasters that could transform into robots were Lodony, Loofa, and Kirk. The other three could transform from a head to a beast and were the only Headmasters to feature this style of transformation—Lione, Trizer, and Shuffler. Takara also made available white versions of these toys, which were given away in stores when you bought a Transformer. All six of these characters were original molds and can sell for hundreds of dollars each. These Transformers were not featured on the television series.

Other exclusive Japanese figures were the six-team merge group of the Trainbots, who could combine together in order to form Raiden. This team was the most popular merge group on the show. They were the only other Diaclone merge group besides the Constructicons to make it into the Transformers Universe. The Trainbots were sold in six individual boxes which came in several different styles, but were also made available to the public in an extremely rare and expensive gift set.

Transformers Headmasters ended with Ultra Magnus being killed off, Galvatron trapped within a block of ice, and Mega Zarak being cut in half by Fortress Maximus. Once the battle was over the Cybertrons left Earth in search of a new planet to inhabit, and to continue the war in outer space. The last scene is of Daniel waving good-bye to the Cybertrons as they fly away into the sunset.

Weirdwolf D-86, $75-90. *Courtesy of Weird Jim Walters.*

Mindwipe can wipe $65-75 from your wallet. *Courtesy of Charles Liu.*

The Destron Targetmasters on average are more popular than the Headmasters. Seen here is Triggerhappy, $140-160. *Courtesy of Charles Liu.*

The Twinbots Pounce and Wingspan. This box originally came with a removable flap which hangs over the window and covers up the figures in the box, $45-55. *Courtesy of Jim Walters.*

The Doucons were robots that transformed into two separate vehicles at the same time, and could only change back to robot mode with both parts present. Battletrap and Flywheels, $25-30 each. *Courtesy of J. E. Alvarez.*

Soundblaster was the re-incarnated version of Soundwave, $455-480+. *Courtesy of Charles Liu.*

Soundblaster is almost identical to Soundwave except...

He could hold two tapes in his expanded chest instead of just one.

A variation is found on the Buzzsaw that came with Soundblaster. He has a higher ridge on his forehead (seen here on the left) than the regular version does.

Soundblaster also came with a figure of Buzzsaw. Loose and complete, they can sell for $175-200.

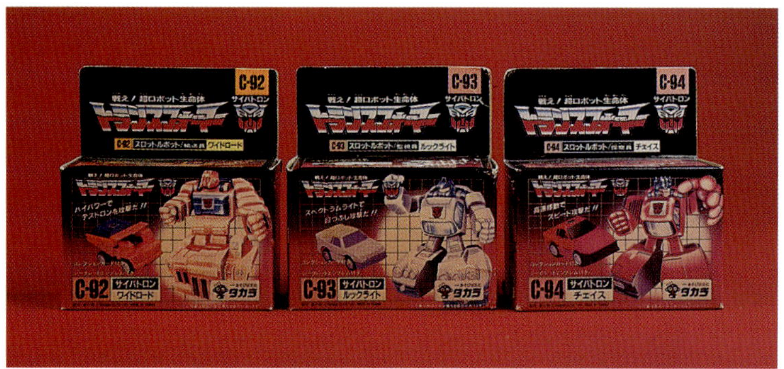

The Throttlebots each sell for about $30-35. *Courtesy of J. E. Alvarez.*

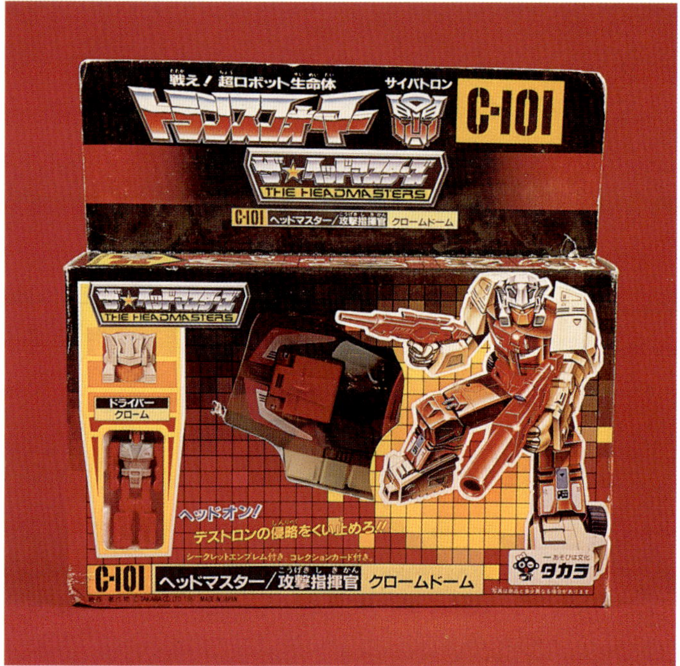

Chromedome was leader of the Headmaster forces, $200-225. *Courtesy of Ron Bahr.*

Artfire is a repainted version of Inferno and is an extremely rare find, $1000+. *Courtesy of Charles Liu.*

Stepper is a black version of Jazz and is another highly desirable piece, $1000+. *Courtesy of Charles Liu.*

The structures of Artfire and Stepper were slightly changed in order to hold a Targetmaster Warrior.

Artfire's fists were retooled a little and Stepper's back piece was changed so they could hold their weapons. Loose they can sell for $300-400+.

Overkill, sealed, can sell for $25-30. *Courtesy of Fran O'Boyle*

Twincast was the reincarnated version of Blaster and is another hard-to-find item, $550-600+. *Courtesy of Charles Liu.*

Loose and complete, Twincast can sell for $225-250.

Like Soundblaster, Twincast also holds two tapes in his chest.

The Steeljaw figure which came with Twincast has a sticker variation on it seen here on the left. When placed inside Twincast, the sticker is decoded by the transparent orange plastic that allows you to see it inside. Soundblaster's Buzzsaw has a similar sticker.

Gurafi and Noizu were a set of Japanese exclusive tapes that are always wanted by collectors, $300-350 apiece. *Courtesy of Charles Liu.*

Here is Gurafi and Noizu as tapes. Even loose, these figures can sell for $125-150 each.

Besides transforming from tapes to animalbots, they could also merge together and form...

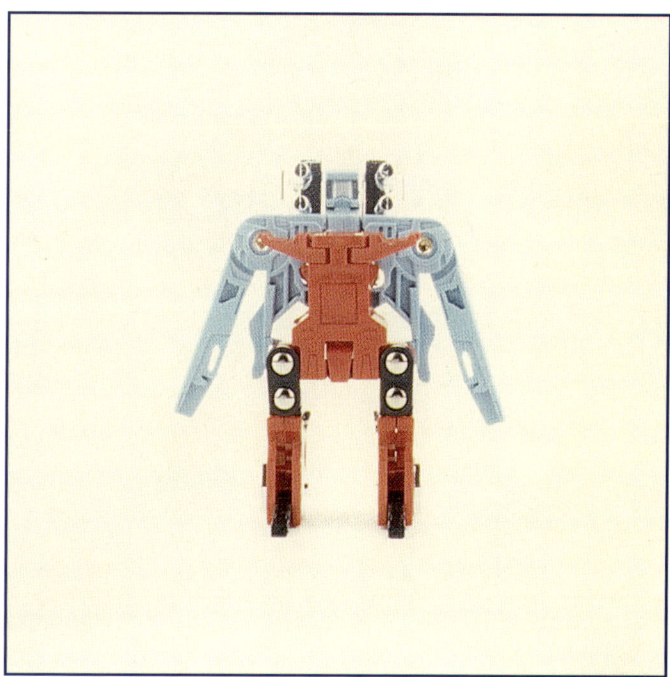

Decibel! This was one of the shots that needed to be retaken because my photographer took the picture of the robot backwards. Thanks Molly.

Here are the first three out of the six Trainbots in their first edition boxes, $150-180+ each. This was also the first picture taken for this book. *Courtesy of George Hubert.*

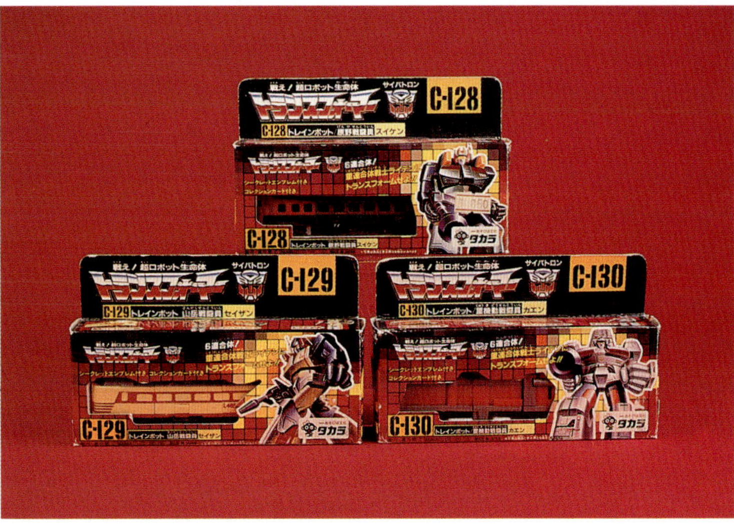

Here are the other three Trainbots which also sell for $150-180+. *Courtesy of George Hubert.*

The Trainbots. *Courtesy of Ron Bahr.*

Besides transforming into individual trains, the Trainbots could all connect together to form one train set through small clips that connect to the bottom of them. They could also fit onto an ordinary train track set.

The Raiden Gift Set is rare and can often sell anywhere between $1200-1800. *Courtesy of Robert T. Yee.*

Together the Trainbots can form Raiden.

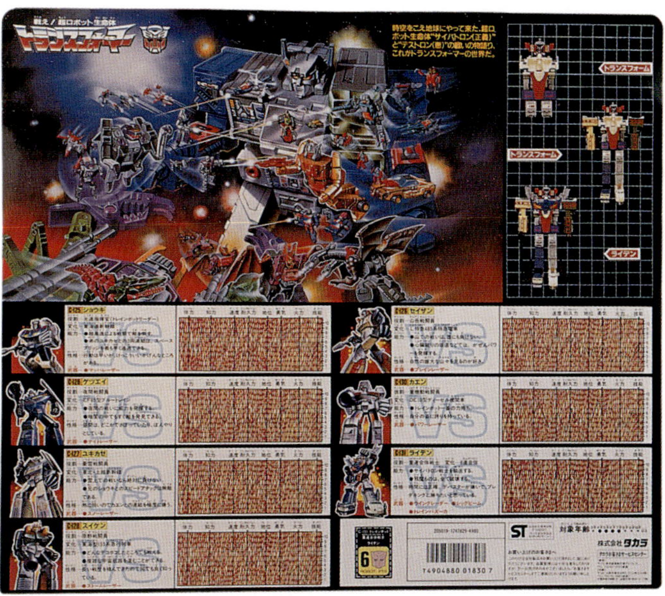

The back of the box shows all six Tech Specs for each Trainbot as well as a Tech Spec for Raiden.

The next series to air in Japan began in 1988 and was called *Transformers Masterforce*. Here the setting was Earth, since the Transformers' home world had been destroyed in the previous TV show. This series revolved around the adventures of the Transformers that had been left behind on Earth, and the story begins ten years after *Transformers Headmasters* ends.

The main characters included the Pretenders, which were robots with removable outer shells that looked like humans or humanoid-like monsters. The leader of the Cybertron Pretenders was Metal Hawk, an exclusive Japanese figure which was the only Pretender to incorporate die cast metal components. The Pretenders could activate their powers and reveal the robot within by yelling out the word "Pretender." Other significant characters included the Jr. Headmasters (Small Headmasters). Now, instead of being smaller robots with larger transforming bodies, these Headmasters were humans outfitted with special suits that would allow them to merge with their transforming vehicles. One very important note about the Jr. Headmaster Minelba—the first female Transformer ever released—was that she was a repaint of the Autobot Nightbeat, who was not female. Being the first female Transformer released makes this toy another piece which is difficult to come by.

As the TV show went on, the Godmasters (Powermasters) were introduced. These were also humans who had exo-suits that enabled them to transform into engines by holding their Godmaster bracelets together and yelling "God on!" Leading the forces of the Cybertron Godmasters was Ginrai (also spelled Jinrai by some), known as Powermaster Optimus Prime outside Japan. Ginrai was only the cab part of the toy; together with the trailer he was Super Ginrai. In Japan, Ginrai was a completely different character than Optimus Prime; he just looked like Optimus Prime. Even the toys differ in many ways. Although the basic bodies remain the same there were several improvements that the Japanese version had over the more widely-released version. First off, the cab of the Japanese figure was made mostly out of die-cast metal and also had clear windows which one could see through. Another change was that the plastic used to create the trailer for Ginrai was a darker gray than Optimus Prime's trailer. The guns used for Hi-Q to sit on in base mode and which double as shoulder cannons were also made out of this darker plastic for the Japanese version. The sides of the legs of Ginrai were painted chrome, rather than remaining the dark blue of the plastic which they are molded from. On the show, Ginrai's trailer served a very different purpose than in the American/European continuity. The original trailer was intended for the original Convoy to merge with, however the trailer was stolen and not recovered until *Masterforce* takes place. This is why only Ginrai could merge with it because the trailer needed a Convoy-like vehicle in order to transform.

There was another Transformer, a Japanese exclusive called God Bomber, that could merge with Super Ginrai in two different ways. The first was by transforming into a trailer that could hook up to the back of Ginrai's trailer in vehicle mode. The other was by merging with Super Ginrai in robot mode to form an even bigger robot named God Ginrai. Both figures were sold individually, but also came in a rare gift set. In this gift set you can find the variation on God Bomber. There is a part on the lower side panels of the chrome portion of God Bomber that doesn't have the tips of two little missiles sticking out of it, which came on the individually released version of this toy.

Other Godmaster items released included extra Godmaster engines named the Powermaster Warriors. The main reason for this was because you couldn't unlock your Godmaster figure without an engine. Unlike the Headmaster Warriors, these extra engines were not new molds, but repaints of the original engines. The characters featured in this set were Boretto, Aquastar, and Zegota. These toys came in small boxes like the Headmaster Warriors and can sell for $50 to $80 apiece.

The Destrons also had several Godmasters. The first two were repaints of Dreadwind and Darkwing, known in Japan as Buster and Hydra. These Transformers had the ability to become robots, jets, and were also able to combine in jet mode to form Dreadwing. Another Godmaster was a black version of Doubledealer who throughout most of the show remains on the side of evil. The final Godmaster was Overlord. This was a robot comprised of a tank and a jet, which also had a base mode. Overlord was unique in that it had two Godmaster engines—one for the tank named Giga, and the other, another female character named Mega, in command of the jet. Overlord is the only Godmaster that can hold its engines (transformed into robots) in a cockpit while in vehicle mode. Overlord was available in Europe, but never in the US or Canada.

Other characters featured in *Masterforce* were new versions of Fortress Maximus and Mega Zarak. This first repaint was a Headmaster known as Grand Maximus, and although it had the same body as Fortress Maximus it incorporated one new feature. This was a Pretender shell made to fit around Cerebros. The new retooled version of Mega Zarak was named Black Zarak. The colors on this Transformer were gold and black while the retooled head and an additional weapon were red. Grand Maximus is the harder-to-get of the two simply because of his sheer size, and in his case size equals a thousand dollars or more depending on condition. Another repainted figure was Sixnight who was formally known as Quick Switch. The Seacons also made appearances, and except for their leader Snaptrap, there were unlimited amounts of each, which were all mindless robots. This was an interesting change for a merge group, which in the past were always limited to five or six members to a team. In Japan, the Seacons would merge into King Poseidon, instead of forming Piranacon. The Destrons were also aided by a little guy named Browning. This was a small robot that only stood a few inches high on the show. Browning was able to change shape into a gun that looked a lot like the original Megatron. On the show, Browning followed Squeezeplay like a loyal dog that never left his master's side.

With the destruction of many of the Destrons, and the defeat of their leader (the god-like being known as Devil Z who controlled the Destrons), the series came to an end in 1989. *Masterforce* featured some of the best animation ever shown on a Transformers television series. Although the show contained many new types of Transformers, once it concluded most of these characters were never seen again.

Metal Hawk was the only Pretender to have die-cast metal parts. *Courtesy of Robert T. Yee.*

A closer look at the inside of the box of this Japanese exclusive figure.

Phoenix and Diver are very common to find in their Japanese packages, $30-35 apiece. *Courtesy of Fran O'Boyle.*

In order to see the figures inside the box without opening it, you could lift up the front lid.

In Japan, Hosehead was named Carb, $50-60. *Courtesy of J. E. Alvarez.*

Minelba was the first female Transformer ever released to the public, $325-350. *Courtesy of Charles Liu.*

Here she is in robot mode. Minelba is highly desired by collectors in many different ways...

Minelba was a repaint of Nightbeat, and loose can sell for $175-180.

Ranger is a repainted version of Joyride, $100-125. *Courtesy of Ron Bahr.*

Hydra and Buster are repaints of the Powermasters Dreadwind and Darkwing, $85-90 each. *Courtesy of Jim Walters.*

In Japan, Powermasters were called Godmasters. To unlock the vehicles so they can transform you need to convert the Godmaster Warrior into an engine and hook it up to its vehicle. The engines were all interchangeable with each vehicle.

Loose, these figures sell for an average of $40-45 each. *Courtesy of Charles Liu.*

Together the jets could form Dreadwing.

Both Hydra and Buster were available in a two-pack. *Courtesy of Charles Liu.*

Overlord could split apart and form two different vehicles. The tank was named Giga and the jet was named Mega.

Overlord was sold in both Europe and Japan, however the European version came in a gold box. *Courtesy of J. E. Alvarez.*

The front of Mega could split off and be its own mini jet.

Godmaster engines Giga (left) and his sister Mega (right) standing in front of a non-transforming vehicle used for Overlords base mode. The Godmasters can only fit into the vehicle by transforming into engines.

Here is a close-up of Overlord when the chest plates are closed.

Overlord was the only double Godmaster ever produced. When you place the engines inside of his chest weapons spring out of his abdomen.

Overlord has one final transformation, which is a base mode. Loose and complete, this figure can sell for $90-110.

Browning was another Transformer taken from a former Takara toy line and is a Japanese exclusive. *Courtesy of Charles Liu.*

On the show, Browning was the size of the actual toy. As a gun he can shoot pellets.

In robot mode, Browning can shoot his fist or a weapon from his right hand (in gun mode, that's where the pellets shoot out from). Browning sells for $60-75 loose.

Blackshadow is a repaint of the Decepticon Pretender Thunderwing, $265-300+. *Courtesy of Jim Walters.*

The back of the Blackshadow box which shows the battle sequence for Transformers Victory.

Bluebaccus was a repaint of the Autobot Pretender Crossblades, $265-300+. *Courtesy of Charles Liu.*

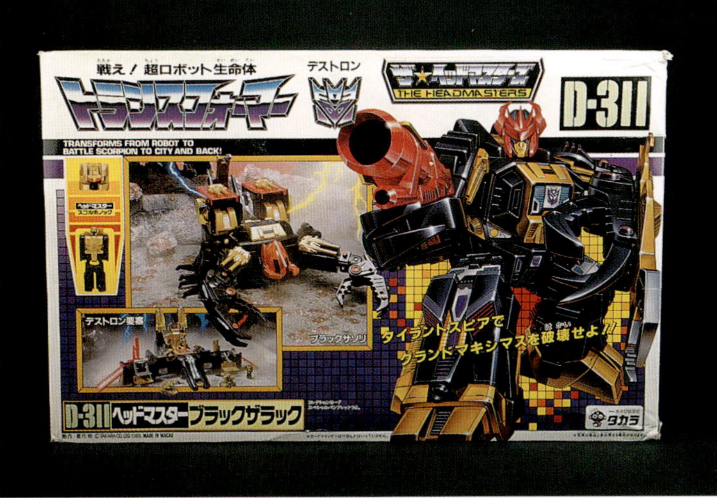

Black Zarak is a repainted and retooled version of Scorponok, $450-550+. *Courtesy of Charles Liu.*

While in beast mode, Black Zarak can move his legs when pushed forward or backwards.

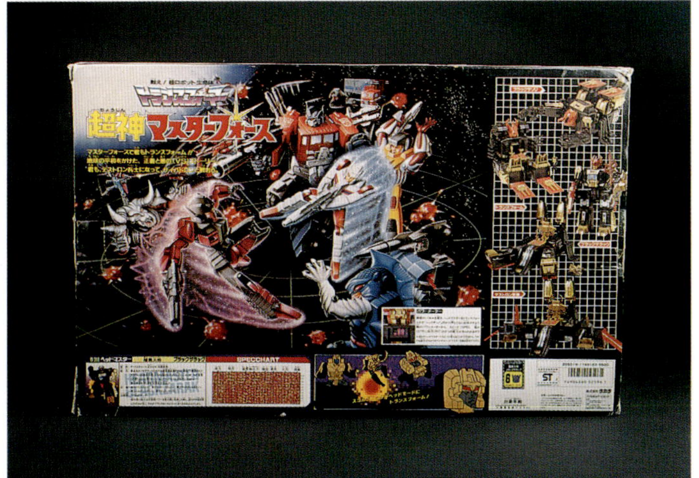

The back of the Black Zarak box.

Base mode.

Side by side the differences between Scorponok and Black Zarak are easier to see. *Courtesy of J. E. Alvarez and Charles Liu.*

Black Zarak's staff was a new addition over Scorponok. His head was also a new design, but could still fit the Headmaster, which was also repainted. Loose, this figure sells for $275-300.

Super Ginrai (Powermaster Optimus Prime) is the combination of the cab and the trailer. Alone the cab is just Ginrai, which was never sold separately, $300-350+. *Courtesy of Robert T. Yee.*

The Godmaster Doubleclouder could transform between a Cybertron and a Destron, $225-255. *Courtesy of Robert T. Yee.*

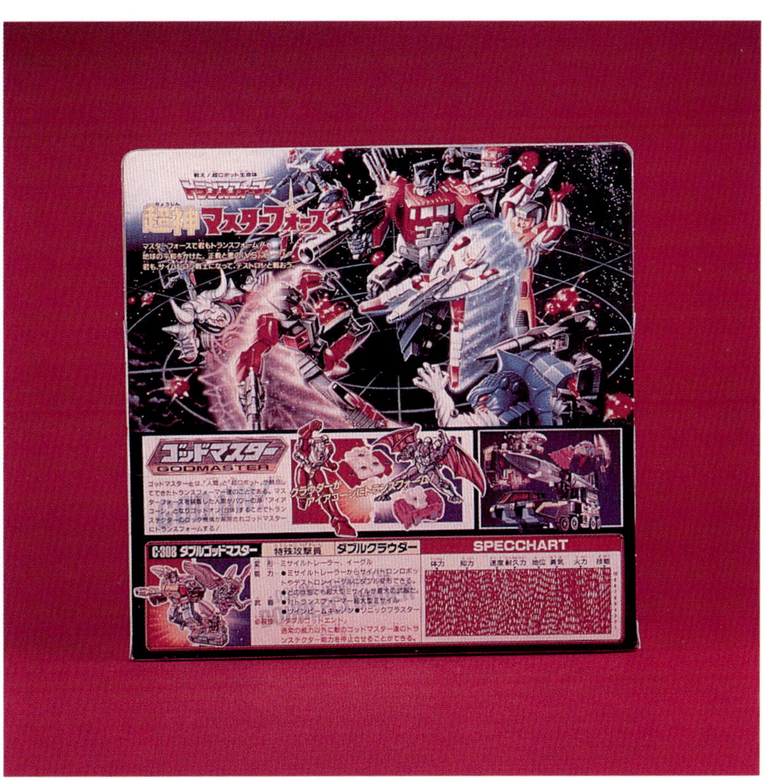

The back of Doubleclouder's box.

This is the same box with the front flap removed. Here you can see the color variation between Doubleclouder and the Powermaster Doubledealer, who is mostly green.

God Bomber could transform into an extension for Super Ginrai in vehicle mode. He also could change into a robot and combine with Super Ginrai to form God Ginrai. $300-350. *Courtesy of George Hubert.*

Below: This is the very hard to find God Ginrai Gift Set which was only available in Japan. In this set is where you find the variation of God Bomber with the missiles sticking out of the front side panels, $450-550+. *Courtesy of Charles Liu.*

Right: The back of the box shows Tech Specs for all three figures and all their different modes.

This is the side view of the box.

This is the Fortress Maximus repaint known as Grand Maximus, who is quite often valued at a grand or more. $1000-1400+. *Courtesy of J. E. Alvarez.*

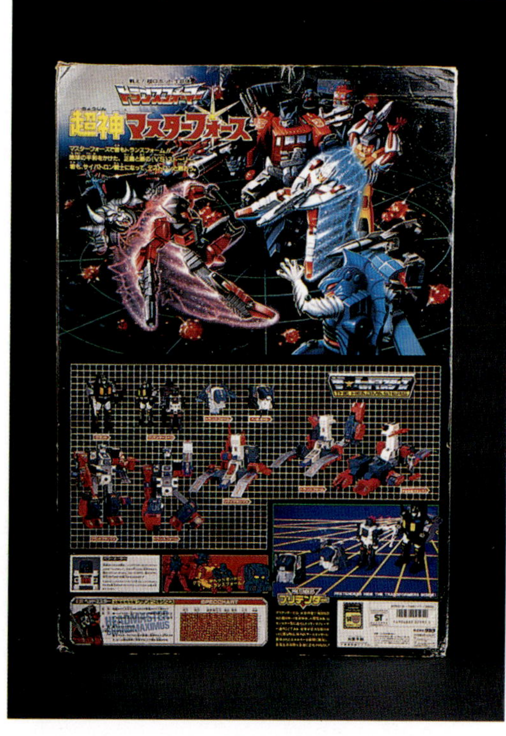

Here is the back of the Grand Maximus box.

On the left is the Godmaster Super Ginrai, while on the right is Powermaster Optimus Prime. *Courtesy of J. E. Alvarez and Charles Liu.*

This is God Ginrai, who is Super Ginrai and God Bomber merged together. $265-300. *Courtesy of Charles Liu.*

Godbomber in robot mode. *Courtesy of Charles Liu.*

A side view of God Ginrai. In this mode he stands about two and a half inches taller than Powermaster Optimus Prime does.

On top is Powermaster Optimus Prime and on the bottom is Super Ginrai.

God Bomber in vehicle mode. The one in the back is the version released individually and the one in front is the gift set version. The variations exist on the side of the chrome portion towards the bottom. The God Bomber from the gift set comes with two missile tips that stick out. The other version doesn't have these missile tips.

Super Ginrai with God Bomber attached to him. *Courtesy of Charles Liu.*

The Headmaster Grand Maximus is highly sought after by Transformer enthusiasts, $550-700+ loose. *Courtesy of Ron Bahr.*

These are the robots that came with Grand Maxx. In the middle is the Pretender shell that can fit Cerebros (shown on the left) inside it. Fortress Maximus did not come with this shell.

Grand Maximus in fortress mode.

Grand Maximus also doubled as a space ship.

With Grand Maxx on the left and Fort Maxx on the right its easy to see how the colors on them were reversed. *Courtesy of Ron Bahr and J. E. Alvarez*

Four out of six Seacons, $45-55 apiece. *Courtesy of Fran O'Boyle.*

This was the day that Ron and I applied the stickers to his Grand Maxx. Notice the content of hair on my arms (and lack of it on my forehead). *Courtesy of Play With This.*

The Japanese (King Poseidon) Seacon Gift Set contained all six characters, unlike the American which only contained five of the figures, $450-525. *Courtesy of Robert T. Yee.*

The next series to be introduced was *Transformers Victory*. The show began in the fall of 1989 on Japanese television and had many great action sequences never before seen on a Transformers series. Some of new characters introduced were the Micro Transformers (the Micromasters), the Brainmasters, and the Breastforce. Most of the toys from this show were never released outside of Japan except for the Micro Transformers and some of the Brainmasters.

The Cybertrons also found a new leader in the form of a Brainmaster named Star Saber. The idea behind the Brainmaster toys was that a small robot would fit into the chest of a transforming vehicle. When the chest was closed the small robot would rise into the head of the robot body and form its face. Other Brainmasters included Laster (also reffered to as Raster), Braver, and Blacker. Another very interesting ability of these three was the power to combine into an even bigger robot named Roadcaesar. These three figures were also sold in Europe, however those versions were not able to merge.

Another popular combiner featured on the show was Landcross, who was made up of Dashtracker, Machattack, and Wingweaver. The individual robots which made up Landcross were made up of two vehicles that were unable to form a another robot by themselves, however each of these vehicles could combine with any other vehicle to form another character. Altogether, there are about thirty combinations which can be forced from these vehicles.

Joining the Cybertrons were Greatshot, who was a retooled version of Sixshot; Galaxy Shuttle, who was able to launch from the Micro Transformers base Countdown; and some of the Micromasters, specifically the Rescue Patrol and God Ginrai. During the course of the show God Ginrai was killed but was later rebuilt by Perceptor, Wheeljack, and Minelba into Victory Leo. Victory Leo could transform from a lion to a robot, and could merge to the back of Star Saber to form a super jet. He also had the ability to combine with Star Saber to form a super powerful robot lifeform known as Victory Saber. Both Star Saber and Victory Leo came packaged separately but were also available together in a gift set.

After the Destruction of Black Zarak and Overlord, a new warrior rose to power. This Destron was known as Deszanrus (also known as Deathsaurus). This character was one of the first Breastforce figures released. Yes, Breastforce! Anyway, the Breastforce figures had a special feature to them, which was the ability to transform their breastplates into an animal and then into a weapon. Deszanrus was the only figure to come with two transforming breast plates. The other Breastforce figures were Leozak, Drillhorn, Hellbat, Gaihawk (also known as Guyhawk), Killbison, and Jaguar (also known as Jallguar) who could all transform and combine into Liokaiser (also known as Liocaesar). Each of these figures changed into a vehicle, and in robot mode their heads took the shape of whatever animal their plate was.

Another merge group which was featured extensively on the show was the Dinoforce. The robots from this group were repaints of the more commonly known Pretender Monsters. Much like the Pretender Monsters, these robots had outer shells only available with the repaints in Japan. These shells were much larger than the Monsters, and made out of a better rubber. Likewise, the Dinoforce figures were also made out of a better plastic. The six robots that could team up to form Dinoking were Gairyu, Kakuryu, Doryu, Yokuryu, Rairyu, and their leader Goryu. On the show Kakuryu, was the comic relief of the group. All were sold individually, and of course were also sold in a very rare gift set.

The show ended with the destruction of Deszanrus at the hands of Victory Saber. After the conclusion of the show, many of the toys continued to be released in Japan in different boxes for the next Transformers TV series *Zone*.

There were six figures that made up Landcross, and there were three sets with two figures in each set. The sets sell for $85-110 each. *Courtesy of Jim Walters.*

The Landcross Gift Set, $325-350. *Courtesy of Robert T. Yee.*

The back of the box shows all the different modes which these six figures can transform into.

The Landcross team had vehicle modes...

You could also add their weapons to them. *Courtesy of Fran O'Boyle.*

Each Landcross figure could merge with any other to form a new robot.

Here is one of the many combinations you can make out of these figures.

And this is another way to combine them.

Not only could these robots merge into Landcross, their weapons could also combine into a mega cannon. *Courtesy of Fran O'Boyle.*

These boxes are from the Chinese release of Landcross. DashTacker C-316, WingWeaver C-317, and MachTackle C-318. *Courtesy of Fran O'Boyle.*

The Roadcaesar Gift Set was made up of Blacker, Laster, and Braver. $250-325+. *Courtesy of Benson Yee.*

Blacker was one of the three Brainmaster robots that make up Roadcaesar, $75-135. *Courtesy of Ron Bahr.*

The back of the box shows how the weapons of each figure could fit onto them in car mode, and then help them form Roadcaesar.

The first edition box of Star Saber is more difficult to find than the second release, $250-275. *Courtesy of Charles Liu.*

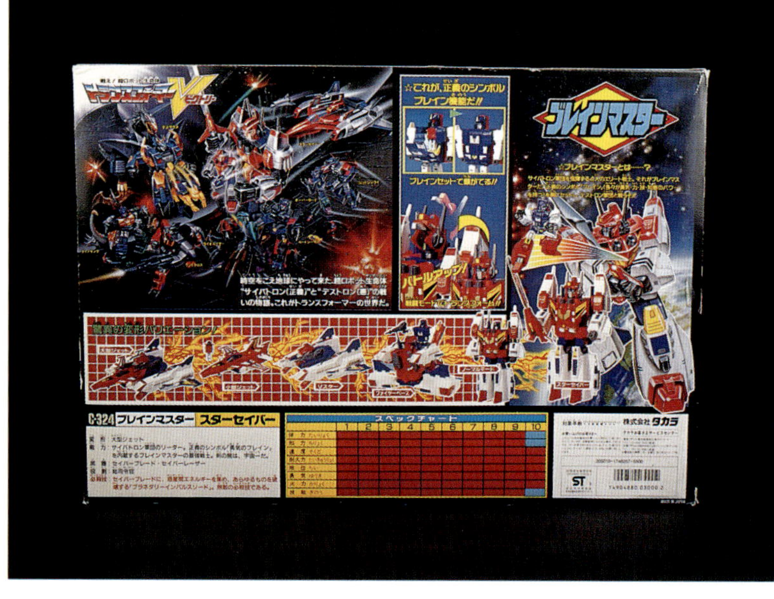

This is the back of the box.

This is the second, more common edition of Star Saber, $200-250. *Courtesy of J. E. Alvarez.*

Star Saber was a Brainmaster, which means that there was a smaller robot that could fit into this robot's chest. When the chest is closed, the smaller robot rises into the larger one's chest, forming its face.

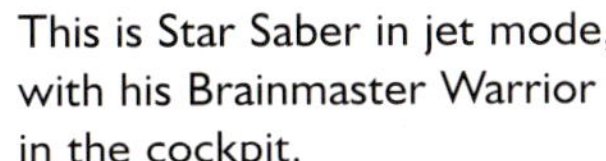

This is Star Saber in jet mode, with his Brainmaster Warrior in the cockpit.

Star Saber could also merge with another jet and become a more powerful vehicle.

This is Star Saber in full jet mode.

When the white and blue jet attachment converts into a robot, the smaller red jet transforms into robot mode to fit into the larger one's chest, while a face plate comes down over the head of the smaller robot to form Star Saber's face. Confused yet? All you need to know is that Star Saber was also leader of the Cybertrons and sells for $145-150 loose. *Courtesy of Fran O'Boyle.*

This is the lesser-known base mode conversion.

Star Saber also combines with Victory Leo to form the super jet.

Greatshot, who fought on the Cybertron side, is a slightly retooled version of Sixshot, who fought for the Destrons. $125-150. *Courtesy of Fran O'Boyle.*

Greatshot...

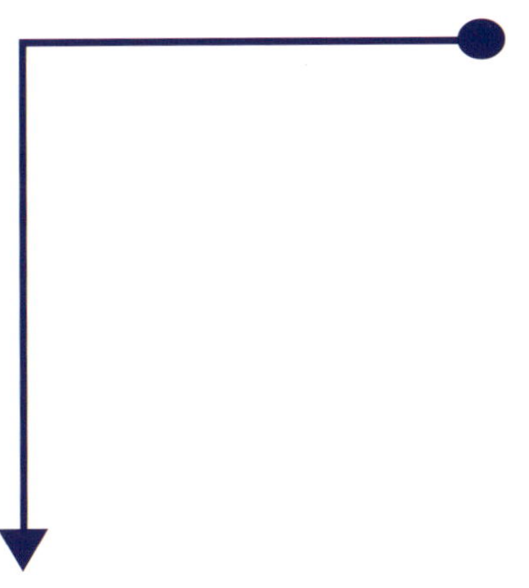

...could...

...transform...

...different...

...six...

...ways! $65-75. Alright, so I stole that from my first book, but that's OK because I used it with Sixshot.

Victory Leo was the reborn version of God Ginrai, $175-225. *Courtesy of Fran O'Boyle.*

Victory Leo sells for $100-125 loose.

Robot mode.

Another rare gift set is that of Star Saber and Victory Leo, $500-600. *Courtesy of Jim Walters.*

The back of the box shows how the two figures in the set transform and team up into a third character.

Together Star Saber and Victory Leo could combine into Victory Saber. *Courtesy of Fran O'Boyle.*

This is the Micro Transformer base Countdown; in Japan it came with the Rescue Patrol. $85-95. *Courtesy of J. E. Alvarez.*

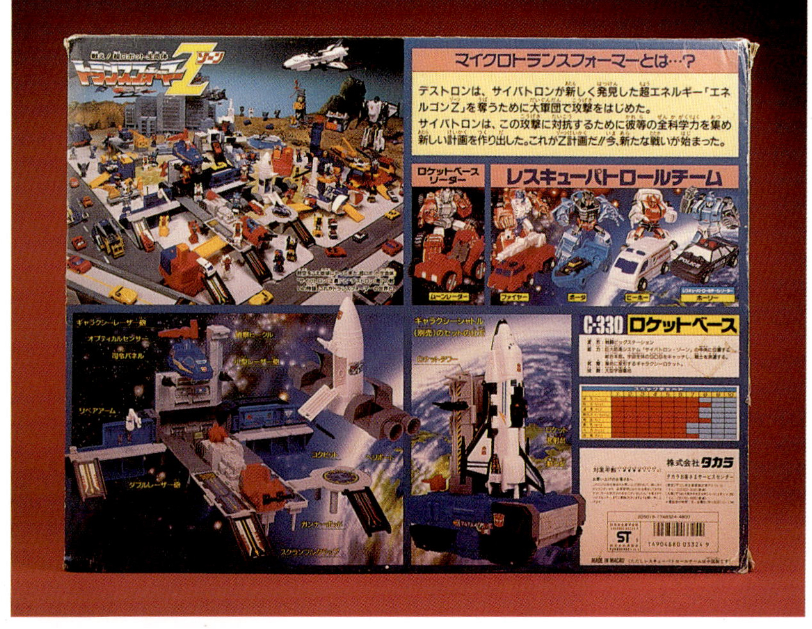

The back of the box shows how Galaxy Shuttle can sit on top of Countdown, ready for blast-off.

The first three out of six figures in the Dino Force (Dinoking) set, $75-85 each. *Courtesy of J. E. Alvarez.*

The six Transformers were repaints of the Pretenders Monsters with new outer shells. Here they are transformed, with their shells, holding on to their weapons.

The other three figures, $75-85. *Courtesy of J. E. Alvarez.*

The six figures transform and merge into Dinoking. *Courtesy of J. E. Alvarez.*

The Dino Force: On top is the leader Goryu. On the second level from right to left are Doryu, Kakuryu, and Gairyu. On the bottom are Yokuryu and Rairyu.

Dinoking on the left and Monstructor on the right. It's very easy to tell these figures apart!

Just like every other combiner team, the Dino Force was sold in a gift set. $450-500. *Courtesy of Jim Walters.*

The back of the gift set.

These are two of the six members that make up Lioceaser. Shown here in their Chinese packaging is Giahawk and Hellbat, $125-140 each. *Courtesy of Fran O'Boyle.*

This is the Japanese Gift Set of Lioceaser, $350-425. *Courtesy of Robert T. Yee.*

Jargua and Killbison. These figures were part of the Breastforce line, $125-140. *Courtesy of Fran O'Boyle.*

The back of the box shows how the Breastforce action works.

Deszanrus was the new leader of the Destron in *Transformers Victory*, $120-130. *Courtesy of J. E. Alvarez.*

This picture shows how beneath the blue and gold breast plate, you could fit another red breast plate. This feature only came on Deszanrus.

Both breast plates could come off and transform into weapons.

Deszanrus in beast mode, along with his two breast plates.

A shot of the rear of the box.

Deszanrus trapped within his box, $250-260+. *Courtesy of J. E. Alvarez.*

Take this Red King!!! *Courtesy of John Marshall.*

The final series intended for Japanese audiences was *Transformers Zone*. Unfortunately only the first episode was ever animated and was released directly to home video. The story went on, however, in a TV magazine in comic strip form. The show also introduced a new line of characters known as the Powered Masters (which are completely different from the Powermasters). These toys were equipped with battery-powered motors that allowed for ramps to move up and down that could launch Micro Transformers or any other small figures.

Victory Saber still had primary control of the Cybertron forces, but now he was aided by the Powered Master Dai-Atlus, who was able to merge with two Powered Masters Sonic Bomber and Roadfire into a base. This base could merge with any Micro Transformer base, as well as combine into a jet known as Big Powered. All three of these figures came packaged individually but also were released in a gift set which is very hard to obtain.

A new enemy appeared in the form of Violenjiger (also reffered to as Violinjaiger). This was a Quintesson-like creature but with only three faces instead of the usual five. This character was like a god, much like Devil Z was in *Masterforce*. In the pilot episode Violenjiger brings back to life nine of the greatest Destron generals in history. Some of these characters included the original merge groups of Bruticus, Abominus, King-Poseidon, Menasor, Predaking, and Devastator. The story line of the show was set up so that these robots could not split apart to form the smaller robots which make them up. The other three generals were Black Zarak, Overlord, and Trypticon, whose appearances were altered so that many of them did not look like they had originally. Instead, they were covered with cloaks, had new weapons, and some even had redesigned faces. Violenjiger resurrected these generals to do his handiwork and capture the Zone energy from the Micro Transformer bases located on the planet Micro.

Once the generals captured the Zone energy from the Micro Transformers, Dai-Atlus and Sonic Bomber flew in to rescue it and kill the generals off yet again. At the end of the episode Dai-Atlus was awarded for his services by Victory Saber.

Once the story was carried on in magazine form, the Cybertrons learned that Violenjiger was able to split apart into three smaller characters, each with a bug-like appearance. Towards the end of the story, the trio was defeated but Violenjiger returned in the form of a huge phoenix comprised of the spirits of the nine Destron generals. Dai-Atlus, Roadfire, and Sonic Bomber combined into Big Powered and destroyed Violenjiger once and for all.

Also introduced in the Zone toy line was Metrotitan, another repainted figure. This was the Destron version of Metroplex. The plastic was a bit cheaper than the original figure, but this version came with an additional ramp that could connect from Metrotitan to any Micro Transformer playset. He also came with a repainted version of the Sky Stalker figure known as Metrobomb. These characters were not featured in the pilot episode.

Galaxy Shuttle was released twice; this is the second edition box. The first box for *Transformers Victory* is harder to come by, $275-350+. *Courtesy of Jim Walters.*

As a shuttle, Galaxy Shuttle had a cockpit that could fit Micro Transformers.

The Micro Transformer Battle Patrol were issued carded, and later on in a box for the *Return of Convoy* line. $15-20. *Courtesy of Zachary Clark.*

Here is the Battle Patrol loose.

The four in robot mode. *Courtesy of Zachary Clark.*

The Micro Transformer Race Car Patrol, $20-25. *Courtesy of J. E. Alvarez.*

On an American card...

This is the Off Road Patrol on a Japanese card...

And on a European card... *All three courtesy of J. E. Alvarez.*

This is Metrotitan a repainted version of Metroplex. The box is smaller than Metroplex and the art work is also different, $260-275. *Courtesy of Jim Walters.*

Metrotitan was also made out of a cheaper plastic than Metroplex, $140-150 loose. *Courtesy of J. E. Alvarez.*

Metrotitan came with all the extra robots that Metroplex did, but he also came with a repainted version of Skystalker (orange and black).

Metrotitan was released for the *Zone* line, but was not in the pilot episode.

Battle station mode.

Another addition to this figure is this black ramp that can connect to any Micro Transformer base.

Which one is Metroplex again? I can't seem to tell them apart!

Flattop kept the same colors as the regular Decepticon version, however in Japan he was a Cybertron. $15-20. *Courtesy of Robert T. Yee.*

Ironworks C-344, $20-25. *Courtesy of Robert T. Yee.*

This Japanese issue of Groundshaker came with a Micro car instead of a jet, like the more common release. $35-40. *Courtesy of Zachary Clark.*

PLACE
STAMP
HERE

SCHIFFER PUBLISHING LTD
4880 LOWER VALLEY ROAD
ATGLEN, PA 19310-9717

This white version of Skyhopper is greatly desired by Micro Transformer fans, $45-55. *Courtesy of Benson Yee.*

A rear view of the box.

The Skyhopper loose.

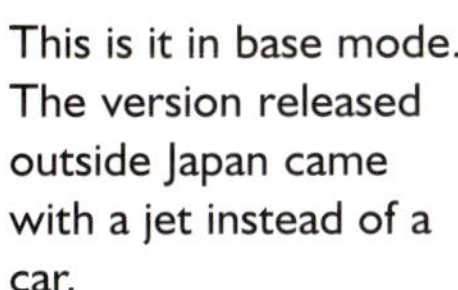

This is it in base mode. The version released outside Japan came with a jet instead of a car.

Sonic Bomber was one of the Powered Masters, who are not to be confused with the Power Masters, $175-250. *Courtesy of J. E. Alvarez.*

This is Diai-Atlus, leader of the Powered Masters and star of *Transformers Zone*. $350-400. *Courtesy of Fran O'Boyle.*

The back of the Dai-Atlus box.

Here is the huge gift set of Roadfire, Sonic Bomber, and Dai-Atlus. $500-650. *Courtesy of Fran O'Boyle.*

The back of the gift set box. All of the Powered Masters were Japanese exclusives.

This shot shows how all three were packaged together in their box. *Courtesy of Jim Walters.*

Sonic Bomber and Roadfire. Roadfire is always the hardest to get out of the set. *Courtesy of Jim Walters.*

Roadfire tank mode.

Sonic Bomber flight mode.

This is Roadfire again, transformed into a base which can connect to any Micro Transformers base.

Dai-Atlus just has this look on his face that says "What! You punks..." He is also battery powered, so he can roll forwards or backwards. *Courtesy of Fran O'Boyle.*

Here are Dai-Atlus and Sonic Bomber in "Zone" mode.

In this mode Dai-Atlus could also roll forwards or backwards.

This is Big Powered, the combination of Roadfire, Sonic Bomber, and their leader Dai-Atlus. *Courtesy of Jim Walters.*

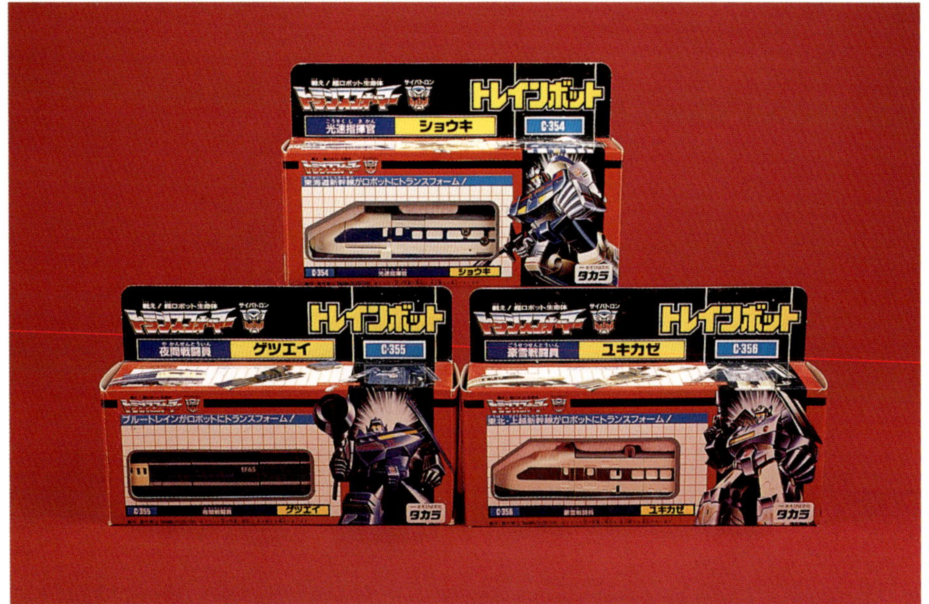

The Trainbots were re-released in harder to find *Transformers Zone* boxes, $165-225 each. *Courtesy of Ron Bahr.*

These are the other three Trainbots, who are also valued at $165-225 each. *Courtesy of Ron Bahr.*

The Transformers saga continued in the pages of Japanese magazines. New story lines and series were also introduced in these magazines which unfortunately were never animated. The first of these series was *Transformers Return of Convoy*, which featured new versions of the original Convoy and Megatron. A Megatron figure for this series was never produced, however a very popular figure of the new reborn Convoy (known as Star Convoy) was. This was a Powered Master figure that came with a very well-made Micro Transformer version of Hot Rodimus. Star Convoy could transform from a truck to a robot, and then into a Micro Transformers base. This figure had the ability to roll forwards and backwards, as well as to roll in robot mode. Other toys from this series included Grandus. This was a robot base for the Micro Transformers, which could turn into a carrier vehicle that hooks up to the back of Star Convoy to be towed by Convoy's mechanized motor. Skygarry was another robot used to transport up to three Micro Transformer launchers (also known as the Micro Trailers). These Micro Trailers could only hold one figure at a time and then launch it out with its spring power. This very hard-to-get figure also transforms into a base or his watch tower mode. He could also combine with Micro Transformer bases. Furthermore, this line introduced something new into the Micro Transformers universe, a character called Six Liner who was made up of six Micro Transformers figures.

Another series was *Transformers: Operation Combiner* (also known as *Mission Combination* or *Operation Combination*). Here the Micro Transformers were aided by repainted versions of Bruticus (called Battle Gaia) and Defensor (called City Guardian). Both these sets are extremely difficult to obtain and run for $600-$850 each. During this line other combiners were released, these however were new Micro Transformer sets consisting of six robots each which could transform from vehicles, to robots, and then merge together into a larger robot. The sets that were released were Six Builder, Six Wing, Six Turbo, and Six Train. This last one was a repaint of Six Liner with a new head. Like the original combiners, these teams used extra pieces that allowed them to connect together. When not in their giant robot mode, these pieces could be linked together to form a Micro Transformers vehicle.

For many years after their demise in America and other parts of the world, Transformers remained a staple in Japanese and some European markets. Transformers did return to Japan with Generation 2, but would not become the acclaimed line it once was until the birth of the *Beast Wars*. **But that's another story...**

Skygarry is an extremely rare toy to find and was issued for the *Return of Convoy* line, $400-500+. *Courtesy of Benson Yee.*

Skygarry came with a Micro Transformer and a launcher that can shoot out any Micromaster, $225-250. *Courtesy of Benson Yee.*

As a jet, Skygarry could extend to fit up to three launchers at a time, as well as hold a figure in his cockpit. There is also a base mode which he transforms into.

These are some of the Micro Transformers that can split apart to form two different robots. They can then transform again and connect to any other Micro Transformer, which splits apart to form a new vehicle of your choosing. $20-25. *Courtesy of J. E. Alvarez.*

This was the first ever Micro Transformer set that could combine into a larger robot, old-school style. Sixliner, $250-300. *Courtesy of Fran O'Boyle.*

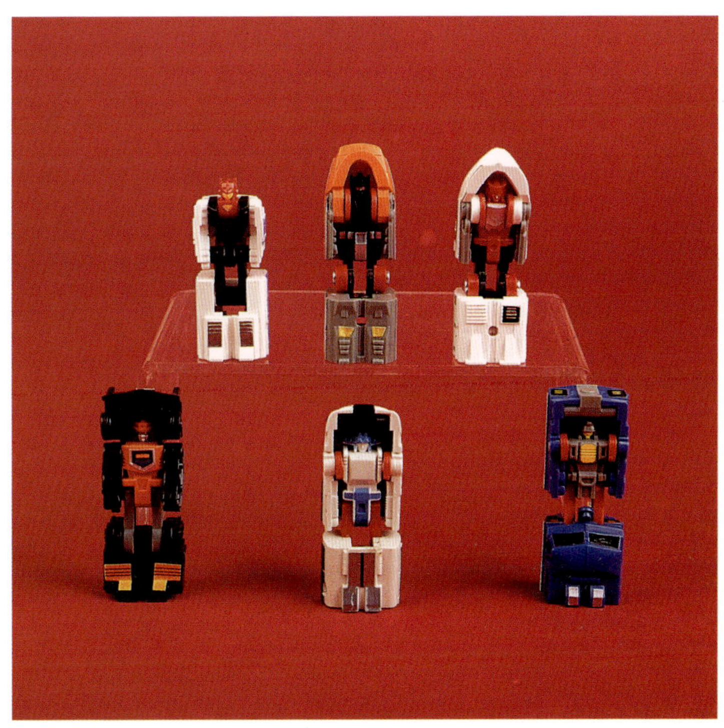

In their robot modes.

In vehicle mode you can hook two of them together at a time to form a mini train.

And together they form Sixliner, who is very sought after by Transfans. *Courtesy of Fran O'Boyle.*

Grandus is another figure highly desired by collectors, $375-425+.
Courtesy of Benson Yee.

The back of the Grandus box

Out of the box Grandus is very impressive, $225-250.

This is Grandus in his base mode. He can connect up to Star Convoy, who is motorized and can power Grandus' lifts.

Another angle of the interior.

In this mode Grandus can hook up to Star Convoy the way that God Bomber hooks up to Super Ginrai.

Star Convoy is the reborn version of the original Convoy, who comes equipped with a Micro Transformers version of Hot Rodimus, $400-450. *Courtesy of Ron Bahr.*

Star Convoy leader of the Cybertrons, $200-235.

Star Convoy was another of the Powered Masters who could roll forward and tow Grandus along, as well as connect to other Micro Transformer bases.

Base mode. Notice Hot Rodimus standing in the center.

Hot Rod and Hot Rodimus.

The Construction and Rescue Patrols, $25-30 each. The Rescue Patrol was first released with Countdown. *Courtesy of Jim Walters.*

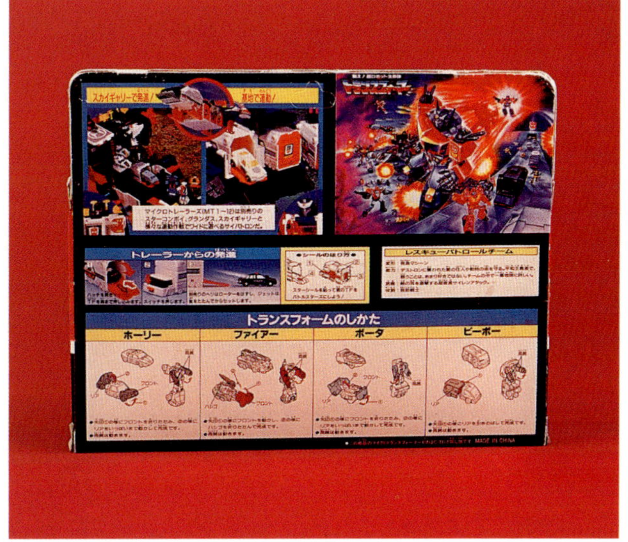

The back of the Rescue Patrol box.

The Battle Patrol first came carded and were later issued in this box, $25-30. *Courtesy of J. E. Alvarez.*

Not many Micro Transformers were issued as Destrons in Japan. That's why so many of them were repainted and released as Cybertrons, as seen here with the Sports Car Patrol, $20-25.

Each set that was released in a box came with a launcher that could hook up to Skygarry. *Courtesy of Zachary Clark.*

The Sports Car Patrol in robot modes.

The team that forms Six Builder. *Courtesy of George Hubert.*

Each Micro Transformer combiner set contained six figures and were all Japanese exclusives.

Six Builder.

The Micro Combiners are pieces treasured by collectors.

Six Builder and Six Wing were issued for *Transformers: Operation Combiner*, $225-250+ each. *Courtesy of George Hubert.*

Here is the Six Wing team.

Six Wing. *Courtesy of George Hubert.*

These are the robots that make up Six Turbo. *Courtesy of George Hubert.*

All together, they form Six Turbo.

In their car robo modes.

Six Turbo and Six Knight, $225-250+. *Courtesy of George Hubert.*

The six robots that make up Six Knight.

Here is how these figures could merge together to form mini train sets.

Six Knight was a repaint of Six Liner, with a new head added. *Courtesy of George Hubert.*

All of the teams had extra parts that would allow them to merge, each of these parts were able to merge themselves and form small vehicles.

These figures came towards the end of the line before it switched over to Beast Wars. The jets are the Predators and the cars are the Turbomasters, $125-145 for each set. These figures were also sold in Europe without being repainted. *Courtesy of Robert T. Yee.*

The Generation 2 Convoy, who is Hero Optimus Prime outside Japan, $65-75. *Courtesy of Robert T. Yee.*

The back of Megatron's box.

Megatron. Instead of D-02 or C-02, the numbering switched over to TRF-2. $65-70. *Courtesy of Jim Walters.*

Hooligan, $15-20. *Courtesy of Jim Walters.*

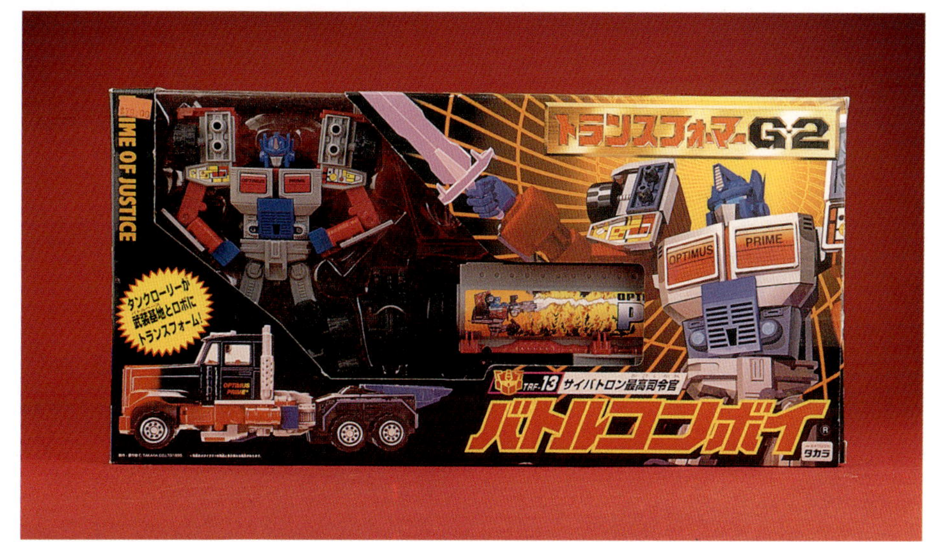

Convoy or "Prime of Justice", $150-165. *Courtesy of Robert T. Yee.*

Dreadwing and Smokescreen, $65-75. *Courtesy of Robert T. Yee.*

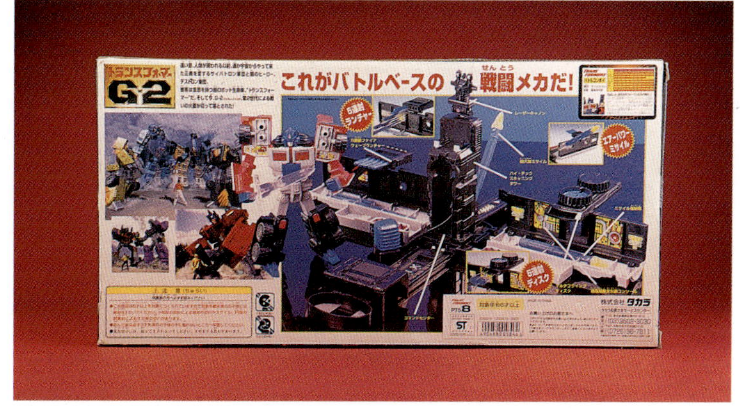

The back of Convoy's box.

Strafe and Jetfire, $15-20 each. *Courtesy of Jim Walters.*

Looks like these might be Chinese versions of the Generation 2 Go Bots. *Courtesy of Robert T. Yee.*

CHAPTER THREE
Other International Transformers

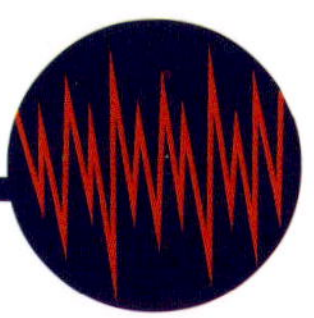

Once Transformers became a fix in America and Japan, toy companies had their eyes set on European markets. In Europe there are three main companies which distributed Transformers—Milton Bradley, Gigi, and Hasbro.

Milton Bradley only distributed Transformers during the line's first two years. The packaging remains almost identical to the American version, only many of the boxes had English, French, German, and Spanish on them. The toys themselves remained the same. Once Hasbro stepped in to distribute the toys in Europe, the Milton Bradley logo was taken off the boxes and replaced with Hasbro's. During the early part of the 1980s, in Italy and parts of France, a company named Gigi was responsible for the distribution of Transformer products. Much like the other European releases, the boxes remained fairly similar to the American designs. Once again Hasbro stepped in and began releasing Transformers all over Europe.

Sometime around the early 1990s, Hasbro re-released many of its classic figures in gold boxes which were almost exclusive to Spain. These gold boxes incorporated a lot of new artwork on the back. Some of the classic figures re-released were Optimus Prime, Prowl, Jazz, Sideswipe, Wheeljack, Red Alert, the Protectobots, the Stunticons, the Combaticons, the Aerialbots, some of the Triplechangers, and some of the Dinobots. Many of the combiners which featured metal components had them replaced by plastic ones.

Once Transformers became unprofitable in America the line was canceled. Fortunately in Europe the line went on and introduced new figures which were not available anywhere else in the world. Some of these European exclusives included many new Action Masters. Six new carded figures were released, including Tracks, Sideswipe, Charger, Bombshell, Take-Off, and Powerflash. Each Action Master had a companion that could transform into their helmet or into a backpack. Other Action Masters included the Exo-suits with Circuit and Thundercracker who each came with vehicles that become body armor for the Action Masters. Thundercracker was a repaint of an Action Master Starscream figure. Two more Exo-suits were released which were motorized. These were Rumbler and Slicer. The last four Action Masters to be released were the Elites. These were Action Masters that could transform into vehicles. They were Omega Spreem, Turbo Master, Double Punch, and Lightspeed. All of these figures are very difficult to come by and highly desired by collectors.

Other European toys include the Autobot Turbomasters. These were simple cars that had spring loaded weapons. The Turbomasters included Flash (known as Spin Road in Japan), Scorch (Fire Road), Boss (Mach Road), and Hurricane (Checker Road). Storm was a helicopter Turbomaster and Thunder Clash was their leader. Thunder Clash was repainted and released in America as Machine Wars Optimus Prime. Storm was also repainted and released as Sandstorm. The Turbomasters were also released in Japan in two-packs with the Decepticon Predator jets. These jets could change into robot modes as well as connect to the larger Predators Stalker and Crash to display a little battle scene through a special scope. Stalker and Crash eventually became part of the Machine Wars line as Soundwave and Starscream. The other Predators included Snare (who was called Flare Jet in Japan), Falcon (Shadow Jet), Skydive (Dark Jet), and Talon (Moon Jet). The versus sets in Japan were of Spin Road and Dark Jet, Checker Road and Moon Jet, Mach Road and Flare Jet, and finally Fire Road and Shadow Jet.

Some repainted figures that came out of Japan to Europe were the Autobot Motorvators. These were repainted versions of the Brainmaster combiner team of Roadcaesar. The Motorvators were Gripper (Blacker in Japan), Lightspeed (Braver), and Flame (Laster). The other main difference between the European and the Japanese versions is that the Motorvators lacked the components that would have allowed them to combine together to form Roadcaesar. Another set of repaints was the Autobot Rescue Force. These were repaints of the Breastforce team that merged to form Liokaiser. Only Leozack, Jargua, Killbison, and Drillhorn were re-released. These figures no longer came with their transforming breast plates nor the ability to transform into Liokaiser. Other European figures included the Decepticon Trakkons, Stormtroopers, and the Autobot Lightformers and Obliterators for both factions.

In America *Transformers Generation 2* came and went like a copy of a John Marshall book. However across the pond *Generation 2* flourished which allowed for many packaging variations and several exclusive figures.

The infamous green Cliffjumper from Argentina, $55-65. *Courtesy of Benson Yee.*

Here is another version of Cliffjumper in white, $55-65. *Courtesy of Benson Yee.*

This is what the back of the card looks like.

Another figure from Argentina is this blue version of Windcharger, $55-65. *Courtesy of Benson Yee.*

A yellow version of the same toy, $55-65. *Courtesy of Benson Yee.*

An orange and red version of Bumble Bee. The orange version is the more sought after and sells for $85-95, while the red one sells for $55-65. *Courtesy of Benson Yee.*

The back of the card.

Robot-Man X es "un salto mas hacia el futuro," $45-55. *Courtesy of Benson Yee.*

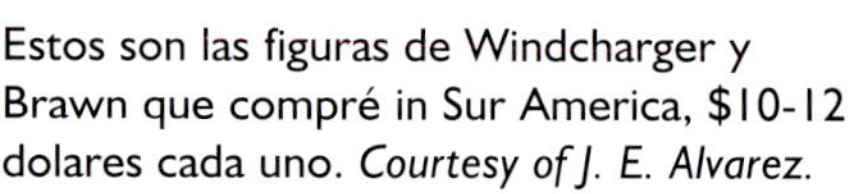
Estos son las figuras de Windcharger y Brawn que compré in Sur America, $10-12 dolares cada uno. *Courtesy of J. E. Alvarez.*

This is the Mexican version of Twin Twist, $20-25. *Courtesy of J. E. Alvarez.*

Ratchet in his European box. Notice that it has the Milton Bradley logo on it. $170-200. *Courtesy of J. E. Alvarez.*

In Europe Tracks was actually released in red, which was his original intended color, $300-350. *Courtesy of Charles Liu.*

Hoist in his European box, $170-200. *Courtesy of J. E. Alvarez.*

Here is the legendary red Tracks, which shouldn't exist according to some collectors. *Courtesy of Charles Liu.*

Hound in his Italian packaging, $165-200. *Courtesy of George Patouhas.*

Chromedome was also issued in Italy. Notice that all the Italian boxes have the GIGI logo on them. $85-90. *Courtesy of Fran O'Boyle.*

And this is the Italian version of Hot Spot, which Fran gave to me on my birthday—the day before the photo shoot for this book, $75-85. *Courtesy of J. E. Alvarez.* Thanks Fran!

These French versions of the Constructicons also featured identifying numbers much like the Japanese versions. $45-55 apiece. *Courtesy of Fran O'Boyle.*

Here is the Canadian issue of Ramhorn and Eject. Both were sold with both gold and silver weapons, however here we have both color weapons put together on the same card. This came out of a case full of the same variation, $75-85+. *Courtesy of Fran O'Boyle.*

Skullcruncher in his Canadian box, $55-60. *Courtesy of Jim Walters.*

Targetmaster Blurr with his Canadian box, $40-45. *Courtesy of Jim Walters.*

Apeface in another Canadian-issued box, $50-55. *Courtesy of Charles Liu.*

Scorponok, Canadian, $150-165. *Courtesy of Zachary Clark.*

The Action Masters were also released in Canada. Here is Rad, $35-40. *Courtesy of Charles Liu.*

Snarl and Kick-Off, $35-40 each. *Courtesy of Jim Walters.*

The Canadian Action Master Optimus Prime, $85-90. *Courtesy of Charles Liu.*

These are the European Autobot Action Master exclusives—Sideswipe, Powerflash, and Tracks. Each of these sell for $20-30 loose each and $75-115+ carded. *Courtesy of George Hubert.*

Many people who haven't seen these figures before may not know that their Action Master companions turn into helmets/backpacks for each of them.

Other European exclusives included the Decepticons Bombshell and Charger. Loose $20-30 and $75-115+ carded. *Courtesy of George Hubert.*

The Action Master Elites were the only Action Masters capable of transforming. Here are Turbo Master and Double Punch. Loose $40-45 and $100-130 carded. *Courtesy of George Hubert.*

Here they are again, alongside Take-Off.

Here they are transformed.

These are the other two European Action Master Elites, Omega Spreem and Windmill. Loose $40-45 and $100-130 carded. *Courtesy of George Hubert.*

Originally Omega Spreem was supposed to be called Omega Supreme, which is why they look so much alike.

Circuit had a car that could transform into an exo-suit around his body. *Courtesy of Benson Yee.*

Thundercracker was the one other figure whose vehicle became an exo-suit, $150-165. *Courtesy of George Hubert.*

Thundercracker was just a repaint of Starscream; the only thing new was the suit.

This is Slicer, one of two figures with motorized Action Master vehicles, $100-135. *Courtesy of Robert T. Yee.*

Rumbler was the other figure with a mechanized car, $100-135. *Courtesy of Benson Yee.*

Slicer came with a repainted Wheeljack, $45-55

The back of Slicer's box: his Tech Specs, a detailed shot of his body, and an Action Master battle sequence.

All that was needed to transform these vehicles was to stand them up...

The Rescue Force figures were repaints of Lioceaser figures, except they lacked the ability to merge into one robot, $75-85. *Courtesy of Robert T. Yee.*

Only four out of the six Lioceaser figures were repainted. The last two are shown here, $75-85. *Courtesy of Robert T. Yee.*

The Stormtrooper Drench could shoot out water and change colors when wet, $75-85. *Courtesy of Jim Walters.*

Two other Stormtroopers with the same capabilities were Aquablast and Rage, $75-85 each. *Courtesy of Robert T. Yee.*

Another group which was available in Europe were the Predators. This is their leader, Falcon, $75-85. *Courtesy of George Patouhas.*

Skydive was another Predator. These figures could link up to the larger Predators and display a battle scene through a small lens, $5-7. *Courtesy of J. E. Alvarez.*

The Trakkon Fearswoop, $60-75. *Courtesy of Jim Walters.*

The back of Fearswoop's box with the figure on top.

Calcar was another Trakkon, $60-75. *Courtesy of Robert T. Yee.*

Opposing the Trakkons were Lightformers such as Deftwing, $60-75. *Courtesy of Robert T. Yee.*

The Skyscorchers Tornado, Terradive, Hawk, and Snipe. $25-35 each. *Courtesy of Jim Walters.*

Thunder Clash was the leader of the Turbomasters, $160-185. *Courtesy of Benson Yee.*

This is what the back of the box looks like.

For some reason Hasbro thought that all little boys wanted to play with neon-colored robots... I'm sure he won't stick out in battle.

This figure was eventually released in America as Machine Wars Optimus Prime.

Thunder Clash's trailer in the tradition of Optimus Prime could transform into an attack base.

The Predator Stalker was one of the figures you could connect to your smaller Predators to reveal a battle scene. $75-85+. *Courtesy of Benson Yee.*

Stalker and Storm were also repainted and released with the Machine Wars line, unfortunately not all of their missiles and weapons were included in that release.

Stalker next to his enemy the Turbomaster Storm.

Turbomaster Storm, $75-85+. *Courtesy of Benson Yee.*

Crash was the other figure that the smaller Predators could join to reveal the special battle scenes, $125-160. *Courtesy of Robert T. Yee.*

The back of the box.

The Obliterator Pyro, $45-65. *Courtesy of Benson Yee.*

The back of his box.

The front part of Pyro transformed into the robot, and the back turned into a small base.

Pyro in vehicle mode.

Clench in his vehicle mode.

Clench was the Decepticon Obliterator.

Here he is one more time inside his box, $45-65. *Courtesy of Benson Yee.*

Es "todo un reto en tus manos" con los del Astro Squad, $35-40. *Courtesy of Fran O'Boyle.*

For more Mexican Transformers, here is the Sky Patrol, $20-25. *Courtesy of Fran O'Boyle.*

The Mexican Micromaster Microcombiner Tanker Truck, $35-40. *Courtesy of Fran O'Boyle.*

The Microcombiner Missile Launcher, $35-40. *Courtesy of Fran O'Boyle.*

This is Fran's favorite, the Groundshaker, $55-60. *Courtesy of Fran O'Boyle.*

The Caurtel General, $55-60. *Courtesy of Fran O'Boyle.*

The Battle Squad is "mas alla de la imaginacion", $35-40. *Courtesy of Fran O'Boyle.*

Flattop, $35-40. *Courtesy of Fran O'Boyle.*

The Decepticon Cannon Transport from Mexico, $35-40. *Courtesy of Fran O'Boyle.*

And this is my favorite Micromaster, the Skyhopper, $55-60. *Courtesy of Fran O'Boyle.*

The Spanish issues of Prowl and Sideswipe, $125-140 each. *Courtesy of George Patouhas.*

The Anti Aircraft Base, $55-60. *Courtesy of Fran O'Boyle.*

Jazz, $125-140. *Courtesy of Jim Walters.*

The Mexican issue of the Sky Stalker, $60-65. *Courtesy of Fran O'Boyle.*

And this is the "classic" version of the gold boxes for Jazz and Wheeljack, $125-140 each. *Courtesy of George Patouhas.*

Everyone should own an Optimus Prime in a Golden Box, especially for $100-150. *Courtesy of George Patouhas.*

Vortex, Swindle, and Brawl each sell for $20-25. *Courtesy of Fran O'Boyle.*

Oslat, more commonly called Onslaught. On these early '90s re-releases any metal parts these combiners had were replaced with plastic, $40-45. *Courtesy of J. E. Alvarez.*

Streetwise and Groove, $20-25 each. *Courtesy of J. E. Alvarez.*

First Aid and Blades, $20-25 each. *Courtesy of J. E. Alvarez.*

Hot Spot, leader of the Protectobots, $40-45. The Aerialbots were also available in these gold packages but are much harder to come by. *Courtesy of Brad Bricker.*

Instead of coming in those triangle packages that were sold in America, in Europe the Generation 2 Hero Optimus Prime came carded. *Courtesy of Robert T. Yee.*

Ramjet, $55-60. *Courtesy of Robert T. Yee.*

Slag on a Euro card, $55-65. *Courtesy of George Patouhas.*

Starscream also came carded in Europe, $70-75. *Courtesy of George Patouhas.*

Grimlock, $55-65. *Courtesy of George Patouhas.*

Bulletbike is one of the new Powermasters (no relation to the Godmasters/Powermasters of G1) and was released in Europe ($35-40, on the right) as well as in America ($65-75, on the left) which is the harder to get out of the two. Not too many people know about the American G2 Powermaster releases. *Courtesy of Robert T. Yee.*

Meanstreak and Ironhide each sell for $35-40. *Courtesy of Robert T. Yee.*

Staxx on the Euro card sells for $35-40 (on the left) and on the American card for $65-75 (on the right). *Courtesy of Robert T. Yee.*

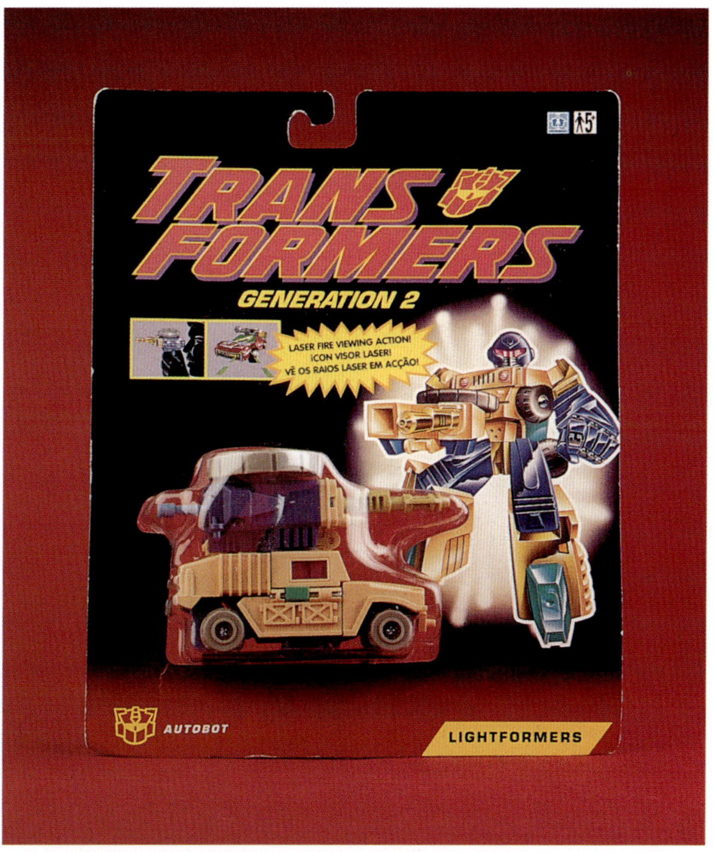

This is a variation of the European card of Ironhide, $40-45. *Courtesy of Robert T. Yee.*

The Decepticon Sparkabots, $30-35 each. *Courtesy of Robert T. Yee.*

The two Autobot Sparkabots, $30-35 each. *Courtesy of Robert T. Yee.*

This is a mis-carded figure. Here is a Jetfire on a Hooligan European card. *Courtesy of J. E. Alvarez.*

Clench was also released in Generation 2 packaging, and only slightly repainted, $35-40. *Courtesy of Robert T. Yee.*

Pyro was also put into a Generation 2 box without being repainted, $35-40. *Courtesy of J. E. Alvarez.*

CHAPTER FOUR
More Oddball Stuff!

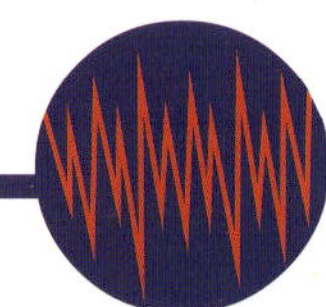

In this chapter we take a look at some of the miscellaneous items that were made with the Transformers name attached to it. Items encompassing this field can reach anywhere from $1 to $100 or more. Although there seems to be a lot of different items, remember that this is only a sample of all the wonderful products out there for us to collect.

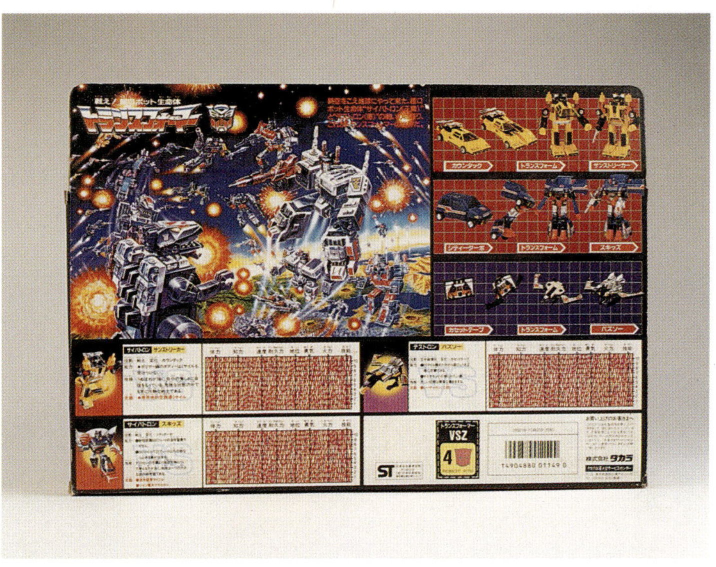

This Gift set was offered as a farewell to the old style of Transformers before the 2010 line took over. On the back of the box is another of the Japanese exclusive battle scenes.

The VSZ Gift Set containing Sunstreaker, Skids, and Buzzsaw is so extremely hard to find that this is the only one I've ever seen, $850-1200+. *Courtesy of Robert T. Yee.*

Here is another of the hardest-to-find sets around. The Good Bye Convoy Gift Set was issued as a tribute to the greatest leader of the Cybertrons, $1000-1400. *Courtesy of Robert T. Yee.*

The Good Bye Megatron Gift Set is perhaps the most desired out of the two Good Byes, $1500-1800. *Courtesy of Fran O'Boyle.*

This is how the figures inside came packaged.

In this set Megatron could also fire pellets and came with his sword, just like the first Japanese issue of this toy.

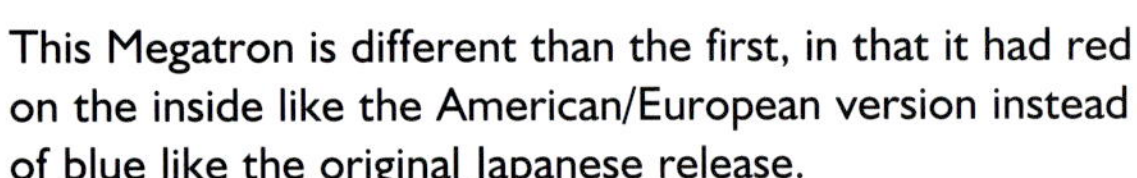

This Megatron is different than the first, in that it had red on the inside like the American/European version instead of blue like the original Japanese release.

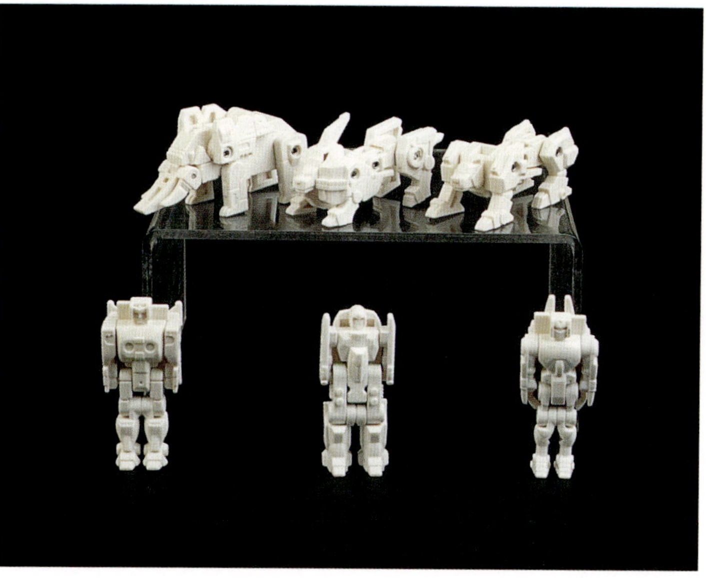

These are the very, very, very rare white Headmaster Warriors in beast and robot modes. *Courtesy of Charles Liu.*

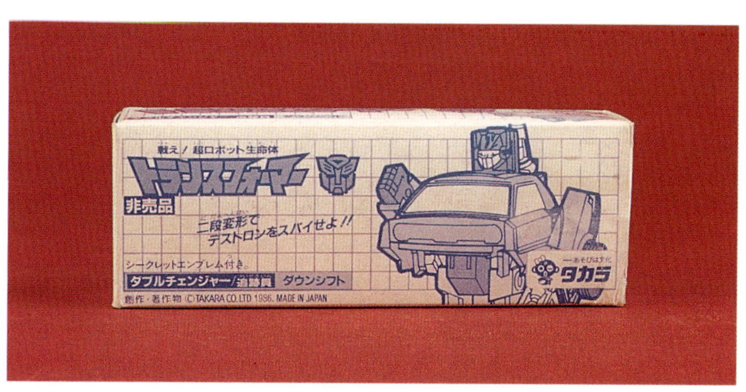

Camshaft was also available in Japan as a mail-away figure, $20-25+.

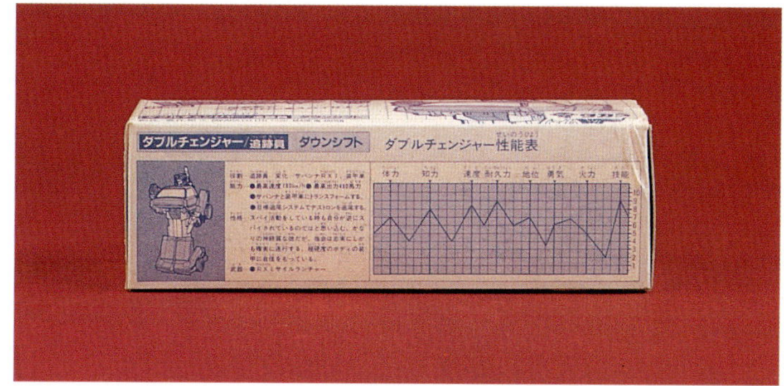

Here are the Tech Specs on the back of the box. *Courtesy of Robert T. Yee.*

The white Headmaster Warriors usually sell on average of over a hundred dollars each.

One of the Godmaster Warriors, $50-80. *Courtesy of Fran O'Boyle.*

Some comic book stores in Japan carried all three of the Godmaster Warriors wrapped together, $135-155. *Courtesy of Robert T. Yee.*

The tape and Rabbicrater in robot mode standing next to it.

Rabbicrater (the small blue Micro Transformer) is C-350, who came with a video of the pilot episode of *Transformers Zone*, $65-75. *Courtesy of Jim Walters.*

This is the Rescue Patrol again, but this time they are packaged with some Legos™ so you can build a little base for them, $35-45. *Courtesy of Jim Walters.*

Takara released a series of Transformers that were based on existing molds and shrunk them down. This line was called the Transformers Jrs. line. Seen here is Convoy, $40-50. *Courtesy of Robert T. Yee.*

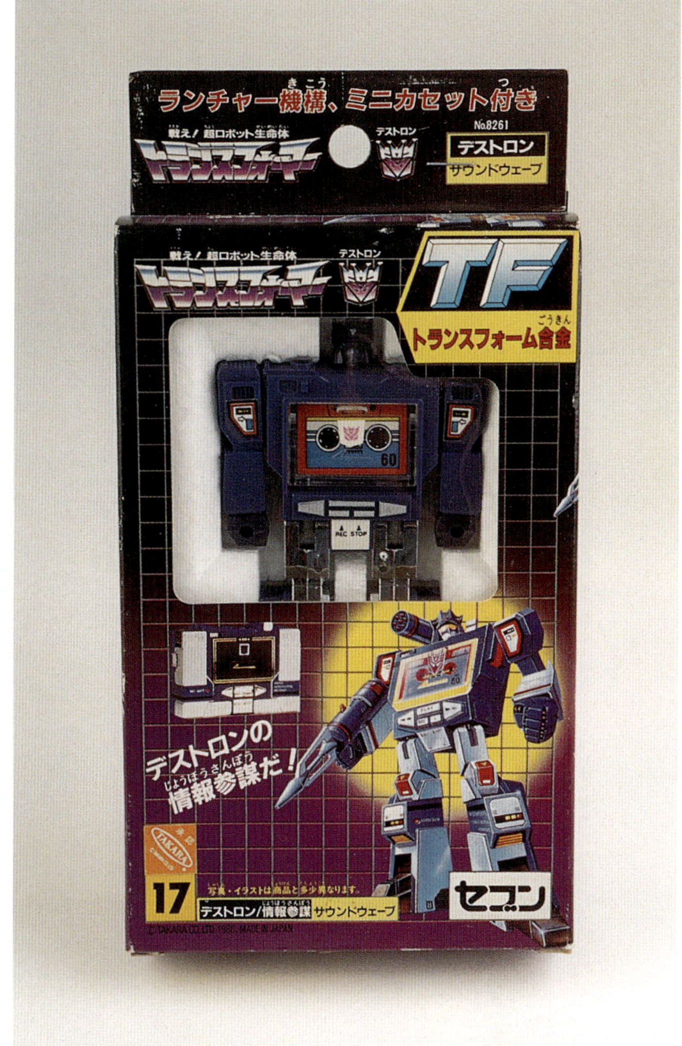

Soundwave was also issued as a Jr., $50-60. *Courtesy of Robert T. Yee.*

Standing next to each other are the Transformers Jr. Convoy and the original Convoy.

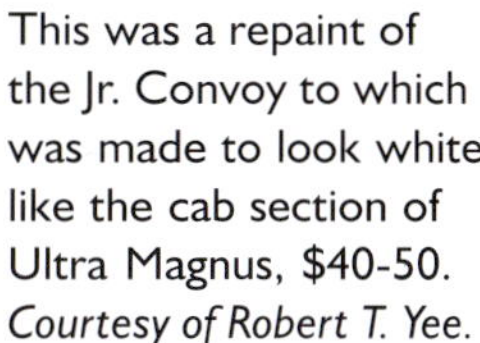

This was a repaint of the Jr. Convoy to which was made to look white like the cab section of Ultra Magnus, $40-50. *Courtesy of Robert T. Yee.*

Chromedome Jr., $45-55. *Courtesy of Robert T. Yee.*

Metroplex Jr., $50-60. *Courtesy of Robert T. Yee.*

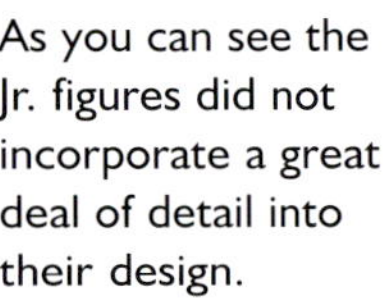

As you can see the Jr. figures did not incorporate a great deal of detail into their design.

Trypticon Jr., $50-60. *Courtesy of Robert T. Yee.*

Metroplex rolling out!

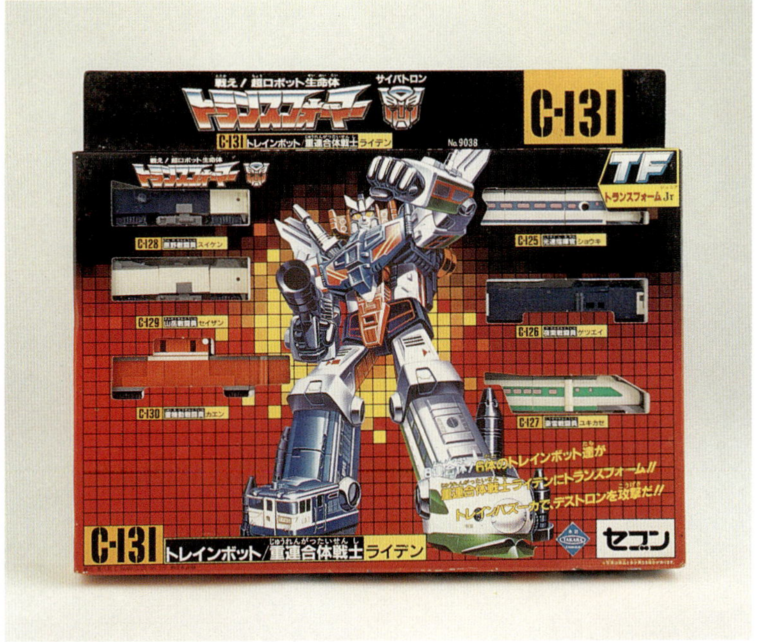

Not many people know about the Raiden Jr. Gift Set. Yes, all six Trainbots, even smaller than the originals, $500-650. *Courtesy of Robert T. Yee.*

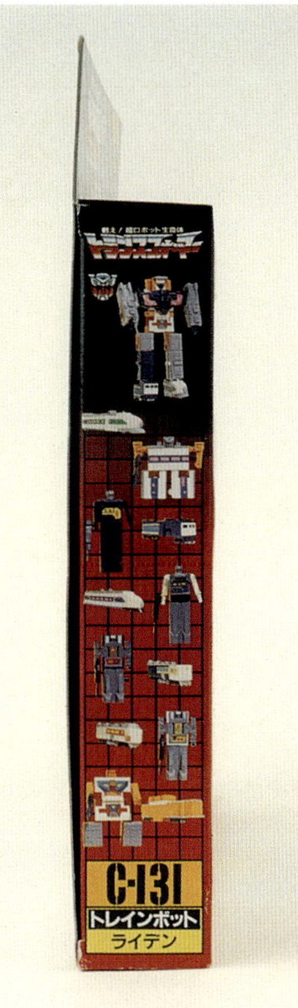

On the side of the box you can see just what each figure looks like.

This is the only way Ginrai was sold individually, $55-60. *Courtesy of Robert T. Yee.*

He also came packaged as Super Ginrai with the option of adding on God Bomber to create God Ginrai, $65-75. *Courtesy of Robert T. Yee.*

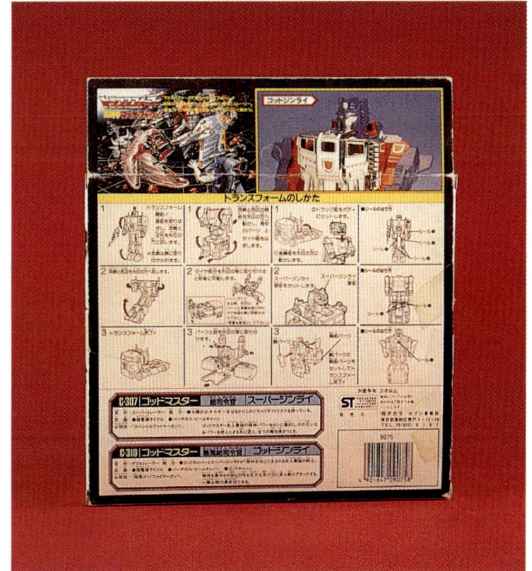

The back of the God Ginrai Jr. box.

From Ginrai...

To Super Ginrai...

To God Ginrai!

Star Saber was also issued as a Transformer Jr. along with a camera that displayed all sorts of things when you looked through it, $85-125. *Courtesy of Robert T. Yee.*

This gift set brings three of the greatest Cybertronian heroes in one box: Chromedome, Convoy, and Fortress Maximus. $75-85. *Courtesy of J. E. Alvarez Congressional Memorial Library.*

Here they are in robot modes. The Convoy is the most detailed out of all of the Transformers Jrs.

This is what they look like transformed. Take note that both Chromedome and Fortress Maximus have Jr. Headmaster Warriors in front of them. George Patouhas gave me that trailer for Convoy because I'm special!

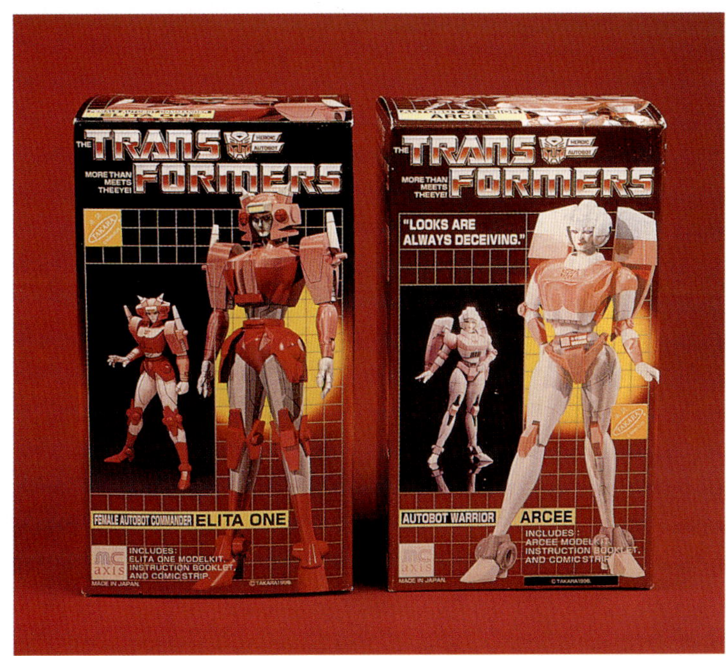

These are the very, very, extremely very rare Elita One and Arcee vinyl model kits. These are both American issues which came out in the early 1990s. *Courtesy of Matthew Camaratta.*

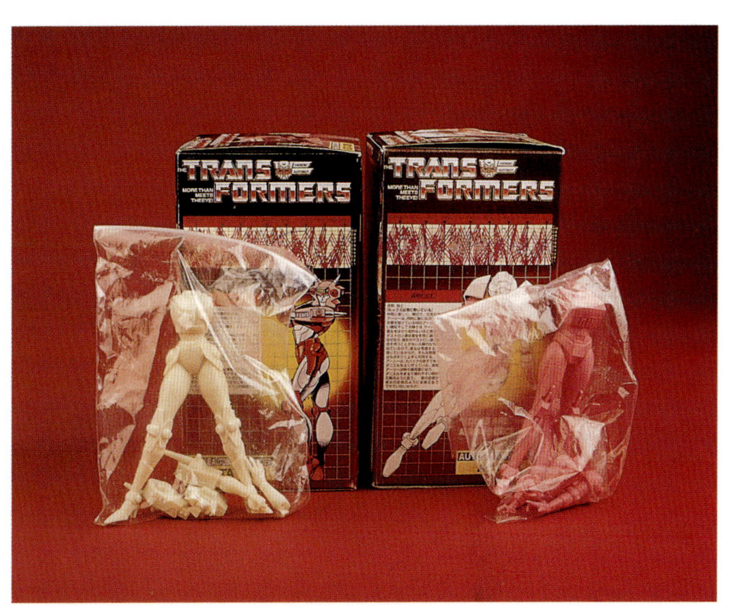

On the back of each box came a Tech Spec for each character. These models stand almost five inches high when assembled, and sell for $500-600 each. *Courtesy of Matthew Camaratta.*

This is the Hot Rod model kit which comes with different attachments you can add on, like his saw from the movie. You can also put a little Matrix of Leadership in his hands. *Courtesy of Robert T. Yee.*

This is what the model kit looks like still sealed in its original bag, $500-600.

There is another rare model kit. No, it's not of Bruticus, but of Battle Gaia—a repaint of Bruticus released for *Transformers: Operation Combiner*, $50-65. *Courtesy of Zachary Clark.*

These models are about two-thirds the size of the actual toys.

These models are very cool because they can transform...

Battle Gaia. *Courtesy of Zachary Clark.*

These are gumball machine model kits of Convoy, Sideswipe, and Starscream. $5-6 a pop. *Courtesy of Zachary Clark.*

One of the model kits built up.

A super deformed model kit of Megatron, $10-15. *Courtesy of Robert T. Yee.*

These models, although very tiny, can still transform.

Here is a side view of the kit.

The Metalforce Convoy, a Japanese toy show exclusive. This 12-inch figure comes with a second head, a Matrix of Leadership, and a second set of hands that can hold his gun. Some of the first figures came with a laser axe which was used by Convoy in the very first episode ever of *Transformers*, $250-350+. *Courtesy of Robert T. Yee.*

Another Japanese toy show exclusive was this approximately 9-inch tall Galvatron.

This was the figure that looked most like Galvatron did on the TV series, $300-325+.
Courtesy of Robert T. Yee.

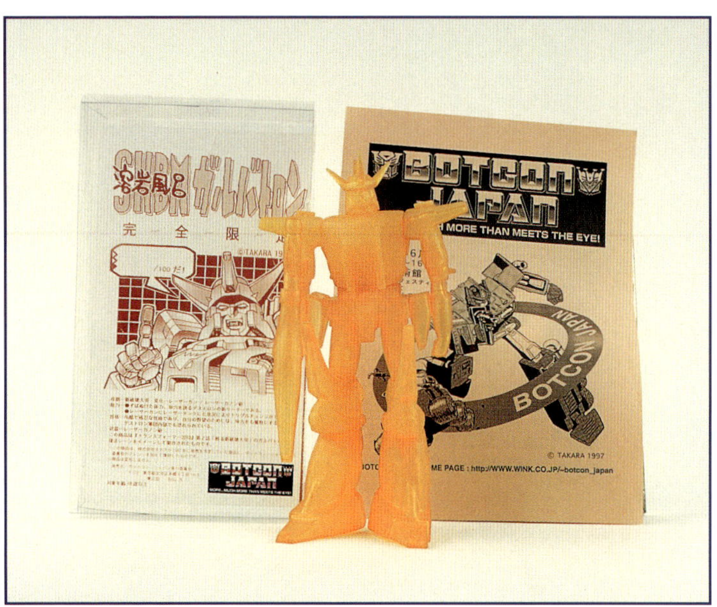

An even more limited version was this orange or "Lava" version, $350-400+. *Courtesy of Robert T. Yee.*

A closer look at the Lava Galvatron figure.

Some smaller vinyl figures of Rodimus Convoy and Galvatron that stand about five inches tall, $125-145. *Courtesy of Robert T. Yee.*

These are the talking vinyl figures of Rodimus Convoy and Galvatron. *Courtesy of Robert T. Yee.*

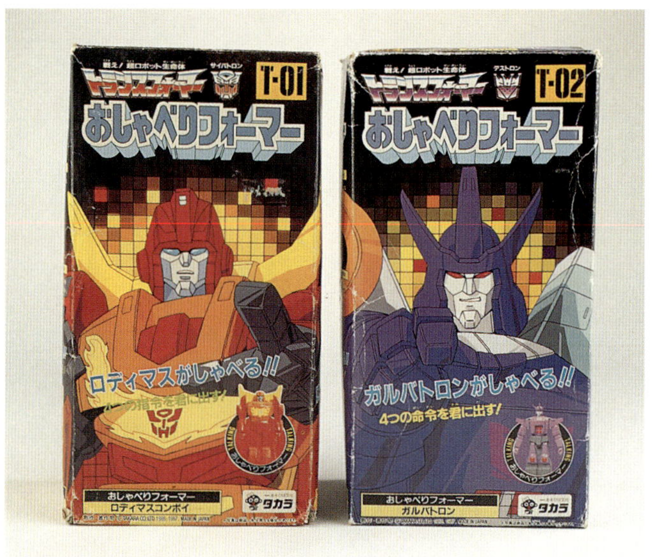

Here are the boxes that they come in. I've seen these figures sell from $125-225 for a single one.

This is one of the many sets of Decoys sold in Japan, $45-55. *Courtesy of Robert T. Yee.*

Here is a larger set containing twenty-two figures, $65-85. *Courtesy of Robert T. Yee.*

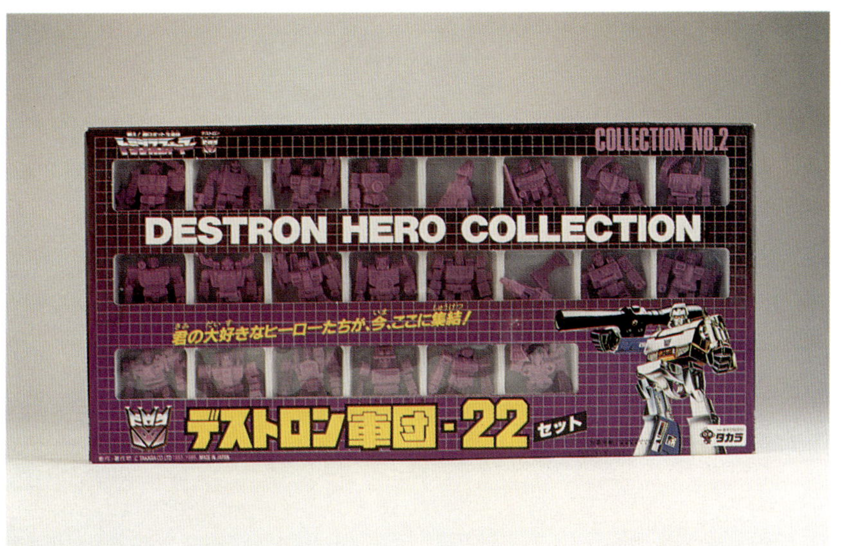

The Destrons were also available in a set of twenty-two, $65-85. *Courtesy of Robert T. Yee.*

The Japanese Transformers board game, $55-70. *Courtesy of Jim Walters.*

This is the harder-to-find Headmasters board game, $65-85. *Courtesy of Jim Walters.*

This board game also came with a small set of Decoys inside it. *Courtesy of George Hubert.*

These are cool items for little kids—replicas of the bracelets used by Siren to activate his powers. I would have loved to have sat in front of the TV as a kid wearing these things while I enjoyed an episode of *Masterforce*, $10-12. *Courtesy of Zachary Clark.*

This is a card set that helped kids learn Chinese symbols, $15-20. *Courtesy of George Hubert.*

A Beast Formers gift set, which included one pure white figure, $200-250. *Courtesy of Robert T. Yee.*

Another card set featuring the Transformers, $8-12. *Courtesy of Robert T. Yee.*

This is the back of the box that shows all the figures that came in the set.

In Japan Battle Beast figures were part of the Headmaster TV series, and thus were sold in conjunction with the Transformers line as Beast Formers. This is their base that transforms into a bird, $85-100.

These are some of the figures which came in pure white. They are some of the harder-to-find Beast Formers items. *Courtesy of Jim Walters.*

Two of the vehicles featuring the Laser Beast Formers have a crystal in their chest instead of a sticker. $75-95 each. *Courtesy of Jim Walters.*

Some of the vehicles used to propel the beasts into battle. $40-55 each. *Courtesy of Jim Walters.*

A rear shot of the box that shows some of the other Laser Beast.

The Beast Formers each came individually packaged, $12-15 each. *Courtesy of George Hubert.*

More Beast Formers... *Courtesy of Jim Walters.*

And three more in the box. *Courtesy of J. E. Alvarez and Fran O'Boyle.*

...and a squid, a snake, and an ape... *Courtesy of Jim Walters.*

There were all different types of Beast Formers—from a duck, to a raccoon, even a sea horse. *Courtesy of Jim Walters.*

There was even a chicken! *Courtesy of Jim Walters.*

Other beasts included a bat and a camel. *Courtesy of Jim Walters.*

These are some of the Laser Beasts which have the crystals in their chest instead of the rubsign stickers. *Courtesy of Jim Walters.*

The Laser Beasts also came individually packaged and are very expensive to obtain. *Courtesy of Jim Walters.*

Here are a few variations which are difficult to obtain. *Courtesy of Jim Walters.*

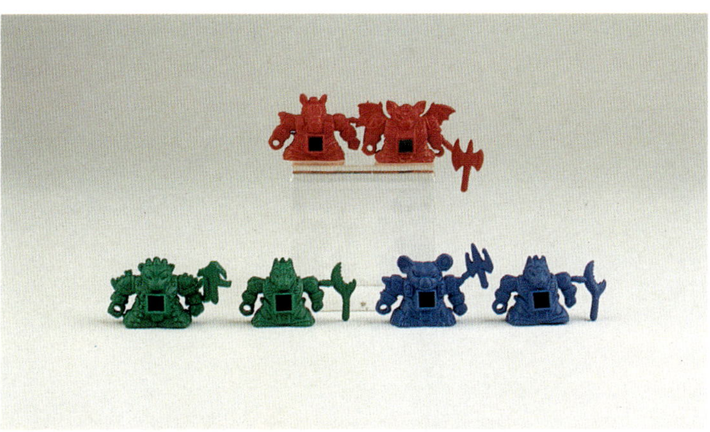

Beast Former pencil toppers. *Courtesy of Jim Walters.*

In Europe Decoys came packaged in these little canisters. *Courtesy of um...um...um..*

Another Venezuelan item is this pop-up book in Spanish, $20-25. *Courtesy of J. E. Alvarez.*

The interior of the book.

This is a reprint of the first issue of the Transformers comic book. I picked this up in Venezuela back in 1989, $10-15. *Courtesy of J. E. Alvarez.*

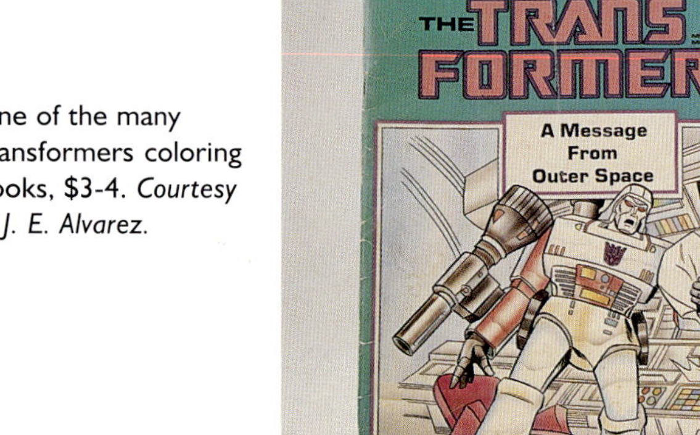

One of the many Transformers coloring books, $3-4. *Courtesy of J. E. Alvarez.*

This is a Sears catalog from 1987. It was lent to me by the Presidents of my fan club, $7-8. *Courtesy of Daniel J. Paglione and Julian P. Paglione.*

Here is the previous book with a soft cover next to another story book with Transformers, $4-5 each. *Courtesy of J. E. Alvarez.*

This is a slightly larger softcover story book, $4-5. *Courtesy of J. E. Alvarez.*

One of the many hardcover books for kids featuring the Transformers, $4-6. *Courtesy of J. E. Alvarez.*

This is the sticker book from the movie, $4-5. *Courtesy of J. E. Alvarez.*

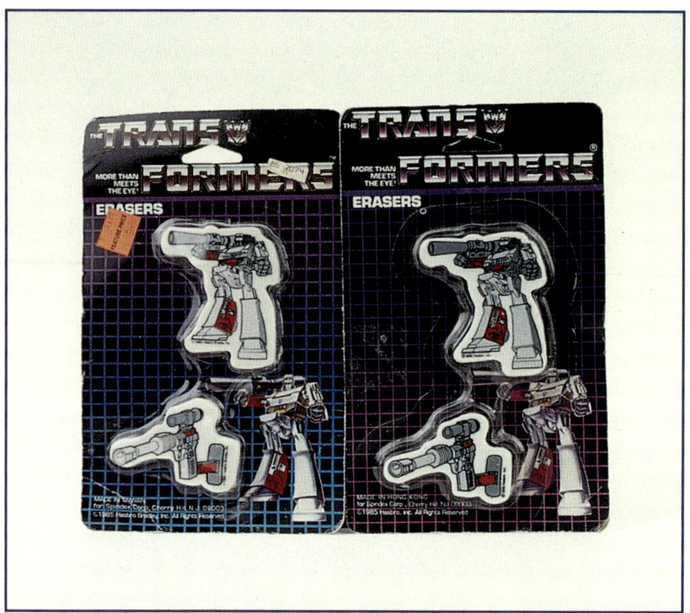

Even on something like the Megatron erasers, you can find color variations on the cards, $5-6 each. *Courtesy of J. E. Alvarez.*

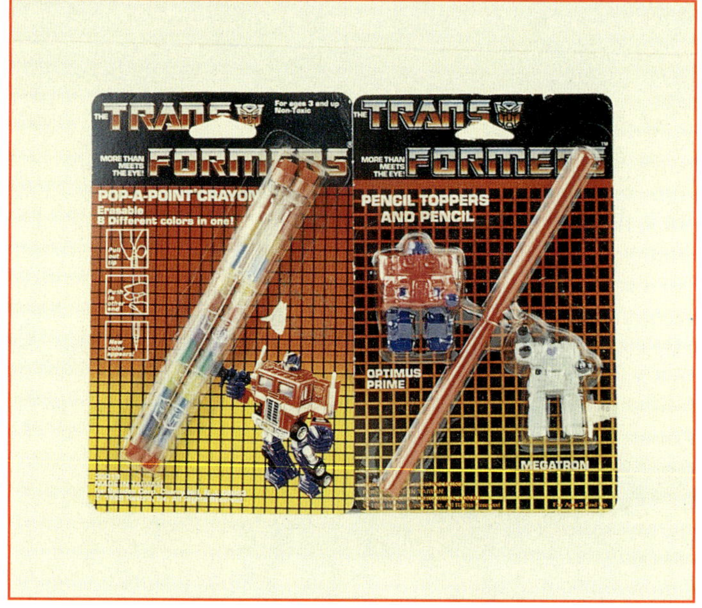

A multi-color pencil set and a set of pencil toppers, $4-6 each. *Courtesy of J. E. Alvarez.*

Unfortunately I have not yet located a variation on the Optimus Prime erasers backer card, $5-6. *Courtesy of J. E. Alvarez.*

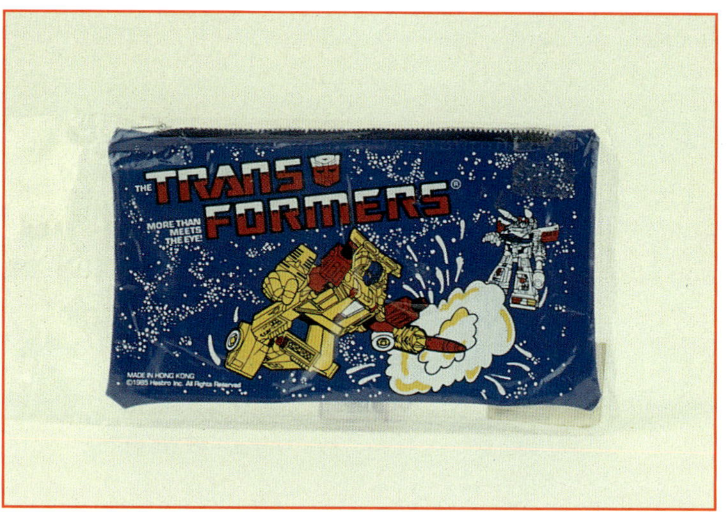

A pencil pouch. A variation actually exists on this item—some have the same picture on both sides and others just have the picture on one side. $4-5. *Courtesy of J. E. Alvarez.*

A carded Starscream stamp sells for $6-8. *Courtesy of J. E. Alvarez.*

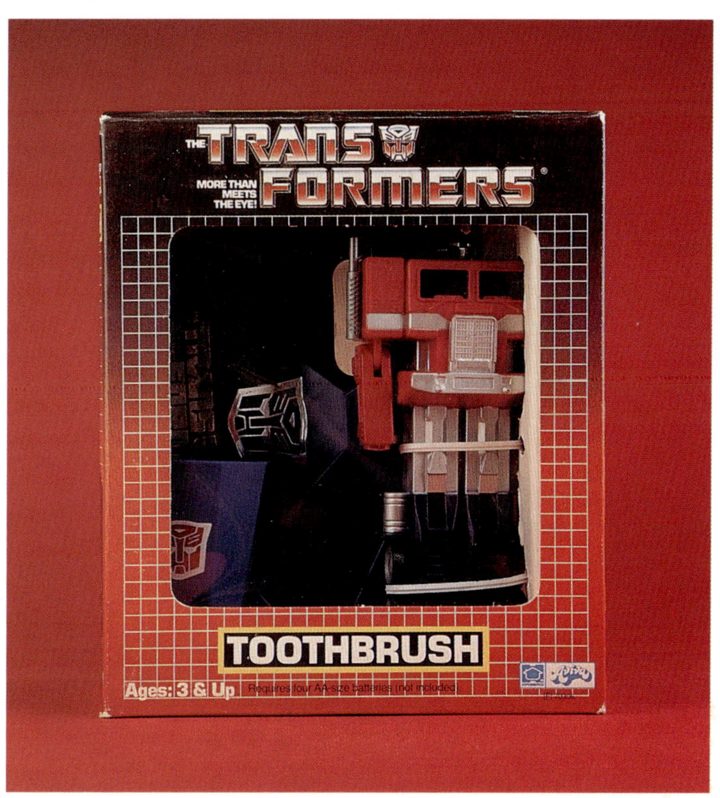

This is something everyone should wake up to do...using Optimus Prime to brush your teeth, priceless. *Courtesy of Robert T. Yee.*

The Electronic Intercom Telephone System! Why bother having a cell phone anymore? $35-40. *Courtesy of Robert T. Yee.*

The Autobot Accessory Pack...leave home without it. $30-35. *Courtesy of Robert T. Yee.*

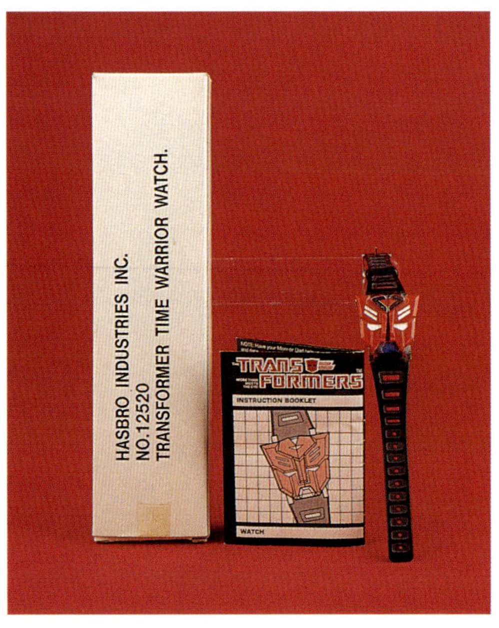

The Time Warrior with its original box and instructions, $60-70. *Courtesy of Robert T. Yee.*

The Tyco Electronic Racing Cars that could actually transform and glow in the dark. $35-40. *Courtesy of J. E. Alvarez.*

The Soundwave tape deck, $15-20. *Courtesy of J. E. Alvarez.*

Reflector in its original mail-away box, $300-350. *Courtesy of Robert T. Yee.*

The mail-away Pepsi Promotional Optimus Prime, $400-450. *Courtesy of J. E. Alvarez.*

Here it is again in robot mode.

One of the prototypes for Scavenger. *Courtesy of Jim Walters.*

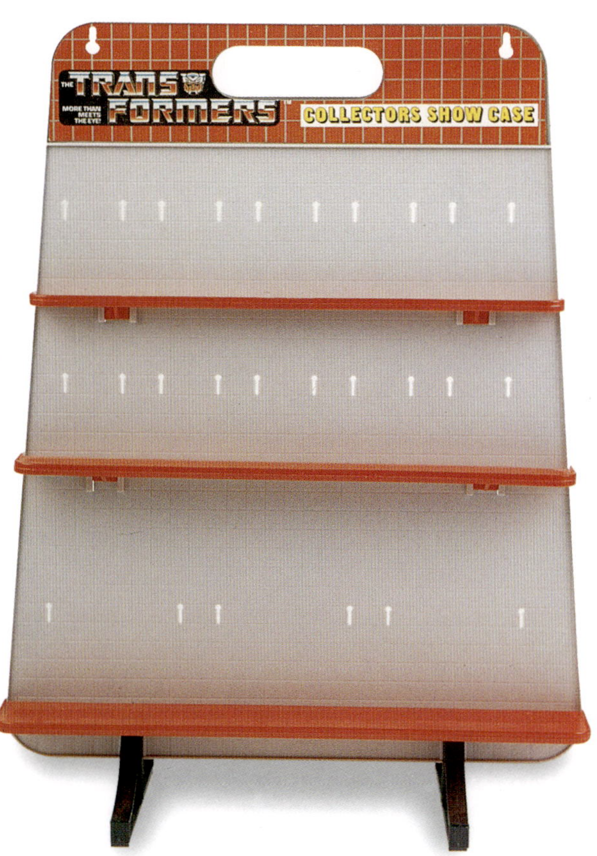

The Collector's Showcase is a little too small for my collection, $15-20. *Courtesy of J. E. Alvarez.*

Someone actually made a Skywarp soap dish...

Your Skywarp soap dish isn't complete without this Optimus Prime shampoo bottle, $15-20. *Courtesy of Robert T. Yee.*

And this is the packaging for it, $20-25. *Courtesy of Robert T. Yee.*

When you take the cab off and stand the bottle upright you have Optimus Prime staring at you while you shower.

A Thrust coin bank, $20-25. *Courtesy of Benson Yee.*

Decorate your room with this Transformers curtain so you can have Optimus Prime stare at you while your sleeping, $8-10. *Courtesy of Julian P. Paglione and Daniel J. Paglione.*

A pillow casing with Optimus Prime on it, $8-10. *Courtesy of Daniel J. Paglione and Julian P. Paglione.*

The Transformers even had a bed tray with them on it, $25-30. *Courtesy of J. E. Alvarez.*

Decepticon and Autobot coffee mugs, $15-20 each. *Courtesy of Will Cintrón.*

The Transformers Paint By Numbers, because to paint by letters is just too complicated, $20-25. *Courtesy of J. E. Alvarez.*

A Transformers lunch box, $20-25. *Courtesy of J. E. Alvarez.*

The Presto Magix removable sticker set, $20-25. *Courtesy of J. E. Alvarez.*

Transformers Shrinky Dinks shrink like magic for $20-25. *Courtesy of J. E. Alvarez.*

Editor and photographer Molly Higgins action figure. She may not transform into anything, but she can shake around in a little grass skirt. Priceless.

Another costume was the Decepto-pack, $35-40. *Courtesy of Robert T. Yee.*

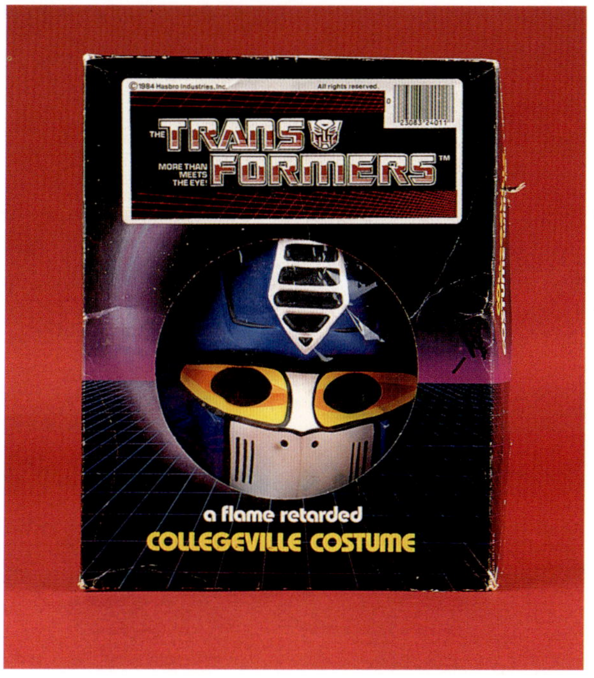

This Optimus Prime costume is flame retarded, $15-20. *Courtesy of J. E. Alvarez.*

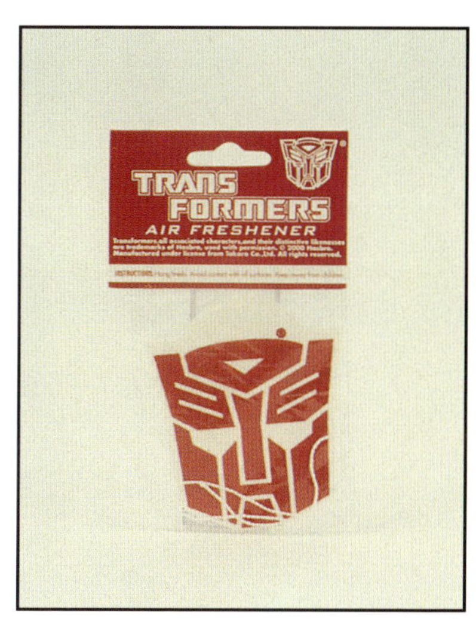

And what Autobot car would be complete without an Autobot symbol air freshener. *Courtesy of J. E. Alvarez.*

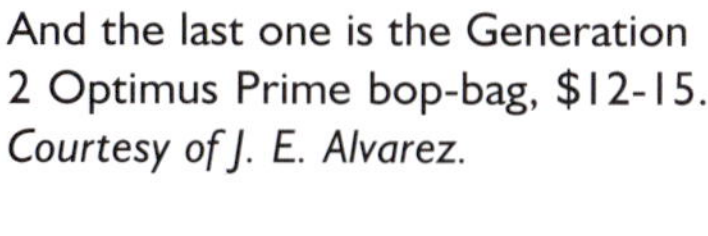

And the last one is the Generation 2 Optimus Prime bop-bag, $12-15. *Courtesy of J. E. Alvarez.*

Alright, everybody, see you in our next installment for the BEAST WARS! Say Good Bye Convoy!!!

To be continued...